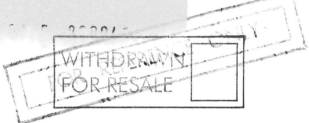

ENVIRONMENTAL HANDBOOK FOR BUILDING AND CIVIL ENGINEERING PROJECTS

CONSTRUCTION PHASE

CONSTRUCTION INDUSTRY RESEARCH
AND INFORMATION ASSOCIATION
6 Storey's Gate
Westminster
London SW1P 3AU
Tel 071-222 8891
Fax 071-222 1708

THOMAS TELFORD SERVICES LTD
Thomas Telford House
1 Heron Quay
London E14 4JD
Tel: 071-987 6999
Fax: 071-538 4101

CIRIA Special Publication 98

Environmental Handbook for Building and Civil Engineering Projects

Volume 2: Construction Phase

Checklists, obligations, good practice and sources of information

Principal authors

Roger Venables – Director, Venables Consultancy Services Ltd

David Housego – Associate Director, Phippen, Randall and Parkes

John Chapman – Associate, Phippen, Randall and Parkes

John Newton – Principal, Ecological and Environmental Services

Pamela Castle – Solicitor and Head of the Environmental Law Group, McKenna & Co

Ann Peirson-Hills – Solicitor, McKenna & Co

The authors have endeavoured to provide guidance to the legislation and the position on other matters as at 1 September 1993 with some minor later additions and amendments to 10 December 1993.

Summary

This Handbook contains information and practical guidance on the environmental issues likely to be encountered at each stage in the tendering and construction phases of a building or civil engineering project. It is aimed at informing construction managers, clients, designers and other consultants, engineers and scientists on their obligations and the opportunities open to them to improve the industry's environmental performance.

Its layout has been designed to be accessible to a wide readership – from those seeking a concise overview or summary of relevant issues and legislation to others seeking more detailed guidance on accepted good practice. References to more-detailed guidance elsewhere are provided whenever possible.

The Handbook also forms a useful checklist and inventory of possible impacts and good practice for anyone considering setting up formal environmental management systems at a corporate or project level.

The main sections comprise:

- Introduction to the key issues involved
- Tendering
- Project planning and contract letting
- Site set up and management including traffic management
- Demolition and site clearance
- Groundworks
- Foundations
- Structural work for building or civil engineering
- Building envelope

- Mechanical and electrical installations and their interface with civil and building work
- Trades: joinery, painting, plastering etc
- Landscaping, reinstatement and habitat restoration or creation
- Site reinstatement, removal of site offices and final clear away
- Handover, and guidance on maintenance and records.

Issues covered include:

- policy and forthcoming legislation
- energy conservation
- resources, waste minimisation and recycling
- pollution and hazardous substances

- transport
- waste disposal
- client and project team commitment.

A list of the main references is provided, together with many useful contact addresses and a checklist based on the contents list for individual readers to use on their projects.

This *Environmental Handbook* is one of a series of related documents and is published concurrently with an Environmental Handbook for Design and Specification of Building and Civil Engineering Projects. Anyone involved in design and construction will need both volumes, though it is stressed that wherever an issue requires identical guidance to be given at the design and construction stages, the two volumes do present the same guidance in essentially the same way.

BREEAM (the Building Research Establishment Environmental Assessment Method) provides a procedure which enables a design or completed building to be assessed on the extent to which it addresses environmental issues. The different versions which currently cover new offices, supermarkets and superstores, existing offices and new homes will each be subject to periodic revision, reflecting progress in what constitutes good practice.

The *BSRIA Environmental Code of Practice for Buildings and Their Services* considers the impact of buildings from the viewpoint of the building services and provides a common strategic framework for the various disciplines involved. A draft document has been piloted and the final version is due for open publication in early 1994. The Code is a working document which makes recommendations on how to minimise the environmental impact of buildings over the entire building life cycle from conception through operation to eventual demolition.

Environmental Handbook for Building and Civil Engineering Projects
Volume 2: Construction Phase

Construction Industry Research and Information Association
Special Publication 98, 1994

Keywords:

Building, civil engineering, construction, environment, energy use, resources, materials, waste, recycling, pollution, handbook, good practice, hazardous substances, internal environment, design use, conservation, environmental legislation, environmental policy.

Reader interest:

Clients, developers, engineers, planners, design and other consultants, builders and contractors, property owners and managers, refurbishment and demolition contractors, Government, local authorities and regulatory bodies, academic and research organisations, legal and other advisors.

© CIRIA 1994
ISBN 0 86017 378 X
Thomas Telford ISBN 0 277 200 1

CLASSIFICATION	
Availability	Unrestricted
Content	Guidance based on good practice
Status	Committee Guided
User	Practitioners in property building and construction

Published by CIRIA, 6 Storey's Gate, Westminster, London SW1P 3AU, in conjunction with Thomas Telford Services Ltd, Thomas Telford House, 1 Heron Quay, London, E14 4JD.

Foreword

This Handbook has been prepared as part of the programme of the Construction Industry Environmental Forum and is aimed at providing practical guidance to engineers, architects, other designers, scientists and managers concerned with the tender and construction phases of a building or civil engineering project. Its preparation is one of many initiatives within the construction industry to improve its environmental performance by identifying and responding to environmental issues in an appropriate manner. It was prompted by an earlier CIRIA project *Environmental issues in construction – A review of issues and initiatives relevant to the construction industry* which concluded that the lack of readily available guidance was a major impediment to improving the industry's environmental performance. This Handbook and a companion volume on design and specification of building and civil engineering projects have been prepared to help the industry by providing background information and good practice guidance on a range of environmental issues.

The objectives in preparing the Handbooks have been:

- to provide a checklist for and guidance on the environmental considerations required at the various stages in the life of construction projects, identifying key activities and decision-making stages which can have significant impact on the environment;
- to produce a framework for the identification of existing information and to provide guidance, at an appropriate level of detail, of current good practice;
- to provide a framework to assist compilation by individual construction-related companies of registers of environmental effects and development of appropriate environmental management procedures, as required in BS7750: 1992: *Specification for Environmental Management Systems*.

The approach has been to provide succinct (rather than encyclopedic) guidance to good practice and obligations and to provide references to the source documents for detailed guidance where they are available. This Handbook addresses the main environmental issues that individual managers, engineers and scientists are likely to face in their day-to-day work in the construction phase, whether they work for the client, designer, main contractor, subcontractors, specialists or manufacturers of components and plant. The aim has been to be practical and to relate the guidance to the realities of working on construction sites, often in remote or congested sites. Ease of use and access to the guidance have also been important considerations.

The extent to which environmental considerations can be built into any project will be influenced by a range of parties including the client, specifier, designer, developer and constructor and is sometimes constrained by cost and time. Whilst this Handbook is not intended as a formal code of practice, it indicates what is presently achievable given the necessary commitment although it is accepted that not all the good practice guidance will be applicable on all projects because of client requirements, budget restraints and other factors.

The document also complements the activities of the Building Research Establishment and the Building Services Research and Information Association. The involvement of BRE and BSRIA in the production of this document has ensured that the guidance provided is broadly compatible.

This Handbook will help many in the construction and related industries to adopt good environmental practice. It must be stressed that there are many areas where practical guidance on actions that individuals can take to ensure good environmental performance is just not available. In such cases, we have sought to identify the nature of the issue and what limited approaches may be appropriate, and to indicate what further work is needed or under way to fill the gap in guidance. CIRIA would welcome feedback from readers on relevant issues so that future editions of the Handbooks and other reports can benefit from the industry's experience.

It must also be stressed that this Handbook has been prepared in 1993 against a rapidly developing industry scene. Whilst much of the guidance will hold good for a long time, readers must recognise the pace of change and take active steps to ensure they are using the latest editions of the quoted references and are up to date on the legislative position. It is recognised that in some important cases – the use of CFCs is a prime example – yesterday's good practice has become today's environmental problem. We all need to keep abreast of technical and economic developments to attain the objective of continuously improving our environmental performance.

The Handbooks have been prepared by a team led by Venables Consultancy:

- Roger Venables and Jean Venables at Venables Consultancy
- David Housego and John Chapman at Phippen, Randall and Parkes, Architects
- John Newton of Ecological and Environmental Services
- Pamela Castle and Ann Peirson-Hills of McKenna & Co, Solicitors, and
- Penny Mills, Book Designer,

with Paul Bartlett and Sandy Halliday representing the Building Research Establishment (BRE) and the Building Services Research and Information Association (BSRIA) as special advisors to the project. The team has endeavoured to summarise the accepted position on the issues covered as at September 1993 with some minor later amendments and additions.

The work has been guided by a Project Steering Group comprising representatives of CIRIA, BRE and BSRIA, contributors and specialists:

- Mr David Lush (Chairman)   Ove Arup Partnership
- Mr David Adler   Representing the Institution of Civil Engineers
- Mr Paul Bartlett   Building Research Establishment
- Mr Peter Charnley   National Westminster Bank Plc
- Mr Christopher Chiverell   Laing Technology Group Ltd
- Mr Michael Gallon   ICI Engineering
- Mr Ron German   Stanhope Properties plc
- Ms Sandy Halliday   Building Services Research and Information Association
- Mr Harry Hosker   Building Design Partnership
- Mr Richard Jarvis   W S Atkins Consultants Ltd, Environmental Division
- Dr Alan Maries   Trafalgar House Technology Ltd
- Mr John Troughton   Department of the Environment, Construction Sponsorship Directorate
- Dr Owen Jenkins   CIRIA (Research Manager for the Project)

The work has been funded though the Construction Industry Environmental Forum with contributions from Department of the Environment, W S Atkins Consultants Ltd, Ove Arup & Partners, Building Design Partnership, ICI Engineering, the Institution of Civil Engineers, Laing Technology Ltd, National Westminster Bank Plc, Stanhope Properties Plc, Trafalgar House Technology, Lothian Regional Council and National House–Building Council.

CIRIA also wishes to express its gratitude to the following additional organisations who provided information and/or participated in a series of consultation meetings held during the course of the project: the Chartered Institute of Building, Chartered Institution of Building Services Engineers, Construction Industry Council, ECD Partnership, Association of Consulting Engineers, Building Employers' Confederation, Essex County Council, Great Mills Retail Ltd, Green Gauge Environmental Consultants, Galliford plc, David Lloyd Jones Associates, Miller Civil Engineers, National Rivers Authority, Oxford Brookes University, Geoffrey Reid Associates, University of Salford, Sheffield Hallam University, the Steel Construction Institute, Richards Moorehead & Laing, Scott Wilson Kirkpatrick & Partners, Sinclair Johnston Consulting Engineers and Taywood Environmental Consultancy.

How to use the Handbook

Introduction

This section on how to use the Handbook provides:

- an explanation of the structure of the Handbook
- instructions on access to the guidance via the contents list or the index, and
- explanations of the terms used
- explanations of the use of references to other documents.

Structure of the Handbook and access to the guidance

The Handbook is primarily structured by the stages in the construction process, from tendering to handover, so that, knowing the stage in the process that you are presently engaged on, you can gain immediate access to the guidance on that stage by turning to the appropriate page given below:

If, on the other hand, you wish to find out what the guide has to say on a particular environmental issue, irrespective of the stage in construction in which the issue may be of concern, turn to the Index on page 140, where you will find references to all pages where the issue is covered.

Each issue is dealt with in a box or table of two main types.

- The legislative position is reviewed in boxes with two main headings: ***Current legal position*** and ***Policy and forthcoming legislation***, each with references to match.
- The present technical position is reviewed in boxes with different headings: ***Background*** and ***Good Practice***, again with references to match. The ***Background*** section provides sufficient information and discussion of the issue for you to understand why the issue may be of concern and/or to explain why the issue is included in the Handbook. The ***Good Practice*** section provides summary guidance on what is accepted good practice or believed to be good practice, on the basis of existing reports, guidance documents and other publications that have been reviewed as part of the project leading to this Handbook.

Explanations of the issue box contents are provided in the diagrams on pages viii and ix. The Issue Titles and, in the technical boxes, the Good Practice Sections of each box are shaded.

References and further reading

The references quoted are of three main types:

- Background References – documents which you may choose to read to amplify the background information given but which are not essential for the full application of the Handbook;
- Good Practice References – documents to which you ought to have ready access, either by ensuring that your office or site has a copy or by having a personal copy for ease of reference, because the Good Practice guidance makes specific reference to them for further detail;
- Legal References.

The main documents in the second and third groups are separately listed in the list of main references on page 128. Contact details of the main publishers and information sources are given in the list of organisations on page 131.

How to use the Handbook – Legal issue boxes

The *Stage in the Construction process* is given in the running header. A Stage sub-heading is given above the first issue box in any new sub-stage.

The *Issue Title* is a topic of sufficient environmental concern that it merits discussion in a separate box.

The *Issue number* has up to four elements:
C identifies the issue as part of the Construction Handbook
First number is the Stage in the Construction process.
Second number is the sub-stage or issue number.
Third number is the issue number when there are sub-headings.

Current legal position is the main heading for information on and discussion of the legislation affecting the issues identified or the stage in design.
The *references to the current legal position* lists the main acts and regulations which you should at least be aware of and/or need to comply with when dealing with the issue concerned.

The *Policy and forthcoming legislation* section sets out in broad terms a range of relevant developments such as government policy statements, developments in the EC and any of the issues that are expected to be subject to legislation in the relatively near future.

Policy references are any documents referred to in the policy section or which provide additional details of the issues raised.

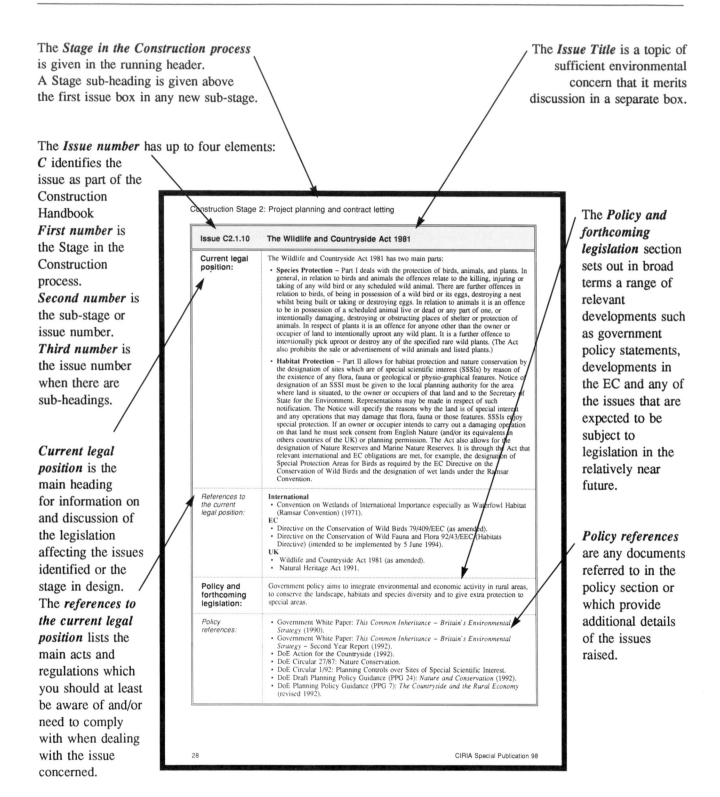

Construction Stage 2: Project planning and contract letting

Issue C2.1.10	The Wildlife and Countryside Act 1981
Current legal position:	The Wildlife and Countryside Act 1981 has two main parts: • **Species Protection** – Part I deals with the protection of birds, animals, and plants. In general, in relation to birds and animals the offences relate to the killing, injuring or taking of any wild bird or any scheduled wild animal. There are further offences in relation to birds, of being in possession of a wild bird or its eggs, destroying a nest whilst being built or taking or destroying eggs. In relation to animals it is an offence to be in possession of a scheduled animal live or dead or any part of one, or intentionally damaging, destroying or obstructing places of shelter or protection of animals. In respect of plants it is an offence for anyone other than the owner or occupier of land to intentionally uproot any wild plant. It is a further offence to intentionally pick uproot or destroy any of the specified rare wild plants. (The Act also prohibits the sale or advertisement of wild animals and listed plants.) • **Habitat Protection** – Part II allows for habitat protection and nature conservation by the designation of sites which are of special scientific interest (SSSIs) by reason of the existence of any flora, fauna or geological or physio-graphical features. Notice of designation of an SSSI must be given to the local planning authority for the area where land is situated, to the owner or occupiers of that land and to the Secretary of State for the Environment. Representations may be made in respect of such notification. The Notice will specify the reasons why the land is of special interest and any operations that may damage that flora, fauna or those features. SSSIs enjoy special protection. If an owner or occupier intends to carry out a damaging operation on that land he must seek consent from English Nature (and/or its equivalents in others countries of the UK) or planning permission. The Act also allows for the designation of Nature Reserves and Marine Nature Reserves. It is through the Act that relevant international and EC obligations are met, for example, the designation of Special Protection Areas for Birds as required by the EC Directive on the Conservation of Wild Birds and the designation of wet lands under the Ramsar Convention.
References to the current legal position:	**International** • Convention on Wetlands of International Importance especially as Waterfowl Habitat (Ramsar Convention) (1971). **EC** • Directive on the Conservation of Wild Birds 79/409/EEC (as amended). • Directive on the Conservation of Wild Fauna and Flora 92/43/EEC (Habitats Directive) (intended to be implemented by 5 June 1994). **UK** • Wildlife and Countryside Act 1981 (as amended). • Natural Heritage Act 1991.
Policy and forthcoming legislation:	Government policy aims to integrate environmental and economic activity in rural areas, to conserve the landscape, habitats and species diversity and to give extra protection to special areas.
Policy references:	• Government White Paper: *This Common Inheritance – Britain's Environmental Strategy* (1990). • Government White Paper: *This Common Inheritance – Britain's Environmental Strategy* – Second Year Report (1992). • DoE Action for the Countryside (1992). • DoE Circular 27/87: Nature Conservation. • DoE Circular 1/92: Planning Controls over Sites of Special Scientific Interest. • DoE Draft Planning Policy Guidance (PPG 24): *Nature and Conservation* (1992). • DoE Planning Policy Guidance (PPG 7): *The Countryside and the Rural Economy* (revised 1992).

28   CIRIA Special Publication 98

Explanation of the legislation issue boxes and the terms used

How to use the Handbook – Technical issue boxes

The *Stage in the Construction process* is given in the running header.
A Stage sub-heading is given above the first issue box in any new sub-stage.

The *Issue number* has up to four elements with the same structure as with the legal issue boxes opposite.

The *Issue Title* is a topic of sufficient environmental concern that it merits discussion in a separate box. It is repeated at the head of the second page of 2-page boxes.

Background is the main heading for information and discussion of the issue for you to understand why it is of concern and/or to explain why it is included in the handbook.

Background references give details of any further reading that you should at least be aware of when dealing with the issue concerned.

The *symbol* ► is used to indicate other sections of this Handbook (C) or its companion volume on Design and specification (D) which also contain relevant information or guidance.

The *Good Practice* section provides summary guidance on what the authors and the Project Steering Group have ascertained is accepted good practice or believe to be good practice. This has been developed on the basis of existing reports, guidance documents and other publications that have been reviewed as part of the project. The extent to which it will be possible to apply the guidance will vary from project to project.

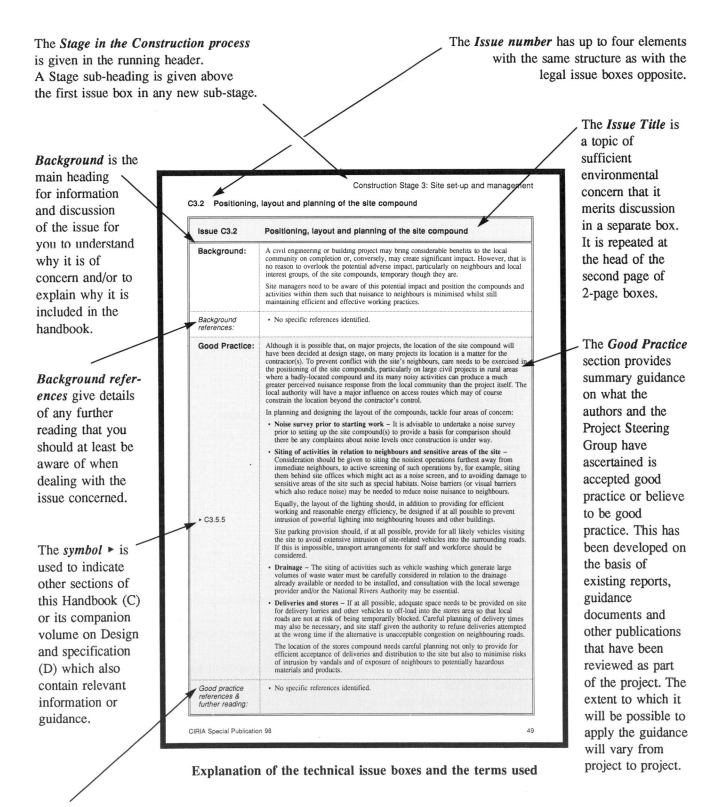

Construction Stage 3: Site set-up and management

C3.2 Positioning, layout and planning of the site compound

Issue C3.2	Positioning, layout and planning of the site compound
Background:	A civil engineering or building project may bring considerable benefits to the local community on completion or, conversely, may create significant impact. However, that is no reason to overlook the potential adverse impact, particularly on neighbours and local interest groups, of the site compounds, temporary though they are.

Site managers need to be aware of this potential impact and position the compounds and activities within them such that nuisance to neighbours is minimised whilst still maintaining efficient and effective working practices. |
| *Background references:* | • No specific references identified. |
| **Good Practice:** | Although it is possible that, on major projects, the location of the site compound will have been decided at design stage, on many projects its location is a matter for the contractor(s). To prevent conflict with the site's neighbours, care needs to be exercised in the positioning of the site compounds, particularly on large civil projects in rural areas where a badly-located compound and its many noisy activities can produce a much greater perceived nuisance response from the local community than the project itself. The local authority will have a major influence on access routes which may of course constrain the location beyond the contractor's control.

In planning and designing the layout of the compounds, tackle four areas of concern:

• **Noise survey prior to starting work** – It is advisable to undertake a noise survey prior to setting up the site compound(s) to provide a basis for comparison should there be any complaints about noise levels once construction is under way.

• **Siting of activities in relation to neighbours and sensitive areas of the site** – Consideration should be given to siting the noisiest operations furthest away from immediate neighbours, to active screening of such operations by, for example, siting them behind site offices which might act as a noise screen, and to avoiding damage to sensitive areas of the site such as special habitats. Noise barriers (or visual barriers which also reduce noise) may be needed to reduce noise nuisance to neighbours.

Equally, the layout of the lighting should, in addition to providing for efficient working and reasonable energy efficiency, be designed if at all possible to prevent intrusion of powerful lighting into neighbouring houses and other buildings.

► C3.5.5 Site parking provision should, if at all possible, provide for all likely vehicles visiting the site to avoid extensive intrusion of site-related vehicles into the surrounding roads. If this is impossible, transport arrangements for staff and workforce should be considered.

• **Drainage** – The siting of activities such as vehicle washing which generate large volumes of waste water must be carefully considered in relation to the drainage already available or needed to be installed, and consultation with the local sewerage provider and/or the National Rivers Authority may be essential.

• **Deliveries and stores** – If at all possible, adequate space needs to be provided on site for delivery lorries and other vehicles to off-load into the stores area so that local roads are not at risk of being temporarily blocked. Careful planning of delivery times may also be necessary, and site staff given the authority to refuse deliveries attempted at the wrong time if the alternative is unacceptable congestion on neighbouring roads.

The location of the stores compound needs careful planning not only to provide for efficient acceptance of deliveries and distribution to the site but also to minimise risks of intrusion by vandals and of exposure of neighbours to potentially hazardous materials and products. |
| *Good practice references & further reading:* | • No specific references identified. |

CIRIA Special Publication 98   49

Explanation of the technical issue boxes and the terms used

Good practice references and further reading may give details of documents needed to secure the detail on the guidance given and/or be suggested further reading if you wish to delve more deeply into current practice on the issue concerned.

Contents

List of abbreviations used

AECB	Association for Environment Conscious Building
AONB	Area of Outstanding Natural Beauty
BATNEEC	Best available technique not entailing excessive cost
BEC	Building Employers' Confederation
BEMS	Building energy management systems
BEPAC	Building Environmental Performance Analysis Club
BRE	Building Research Establishment
BRECSU	Building Research Energy Conservation Support Unit
BREEAM	Building Research Establishment Environmental Assessment Method
BS7750	British Standards Institution *Specification for Environmental Management Systems*
BSI	British Standards Institution
BSRIA	Building Services Research and Information Association
CBI	Confederation of British Industry
CHP	Combined Heat and Power
CHPA	Combined Heat and Power Association
CIBSE	Chartered Institution of Building Services Engineers
CIC	Construction Industry Council
CIEC	Construction Industry Employers' Council
CIEF	Construction Industry Environmental Forum
CIOB	Chartered Institute of Building
CIRIA	Construction Industry Research and Information Association
CFCs	Chlorofluorocarbons
COPA	Control of Pollution Act 1974
COSHH	The Control of Substances Hazardous to Health Regulations, 1988
CPRE	Council for the Protection of Rural England
DoE	Department of the Environment
DoT	Department of Transport
DTI	Department of Trade and Industry
EA	Environmental Assessment
EC	European Community, Environment Council
EDAS	Energy Design Advisory Scheme
EEO	Energy Efficiency Office
EMS	Environmental Management System
EPA	Environmental Protection Act 1990
ESTA	Energy Systems Trade Association
ETSU	Energy Technology Support Unit

FCEC	Federation of Civil Engineering Contractors
HAPM	Housing Association Property Mutual Ltd
HCFCs	Hydrochlorofluorocarbons
HFCs	Hydrofluorocarbons
HMIP	Her Majesty's Inspectorate of Pollution
HMSO	Her Majesty's Stationery Office
HSC	Health and Safety Commission
HSE	Health and Safety Executive
HSWA	The Health and Safety at Work etc Act 1974
HVCA	Heating and Ventilation Contractors Association
ICE	Institution of Civil Engineers
ICRCL	Interdepartmental Committee for the Redevelopment of Contaminated Land
IEE	Institution of Electrical Engineers
IMECHE	Institution of Mechanical Engineers
IWEM	Institution of Water and Environmental Management
NEDO	National Economic Development Office
NEF	National Energy Foundation
NHBC	National House-Building Council
NHER	National Home Energy Rating Scheme
NRA	National Rivers Authority
NRPB	National Radiological Protection Board
OJ	Official Journal of the European Commission
RIBA	Royal Institute of British Architects
RICS	Royal Institution of Chartered Surveyors
RSNC	Royal Society for Nature Conservation
RSPB	Royal Society for the Protection of Birds
SBS	Sick Building Syndrome
SI	Statutory Instrument or Site Investigation
SNCI	Site of Nature Conservation Importance
SP	(CIRIA) Special Publication
SSSI	Site of Special Scientific Interest
TPO	Tree Preservation Order
TRADA	Timber Research and Development Association
UK	United Kingdom (of Great Britain and Northern Ireland)
WWF	World Wide Fund for Nature

Introduction

Environmental issues in construction

The UK Government set out its environmental aims in the White Paper *This Common Inheritance* and, following the 1992 Rio de Janeiro conference on the environment and development, is committed to the production of a Sustainable Development Plan for which a Consultation Paper was issued in July 1993. As well as these Government imperatives, there are also pressures for environmental improvement at European Community, national and local levels, from pressure groups, clients and employees as well as the general public. Wide-ranging changes in attitudes towards the environment are taking place, prompted partly by legislation, partly by client influence in the market place, and partly through personal decisions made by individuals at work and in their day-to-day lives.

The industry's response has been extensive, both in reviewing the key issues it faces and in developing research programmes aimed at producing guidance. The response has included:

- the establishment of the Construction Industry Environmental Forum by CIRIA, BRE and BSRIA;
- research studies by these and other research bodies;
- initiatives led by the Professional Institutions and Trade Associations;
- corporate initiatives.

These initiatives have included:

- participation in the trial use of BS7750: 1992: *Specification for Environmental Management Systems* by construction industry sector groups;
- the development of corporate environmental policies with targets to improve environmental performance;
- BRE's Environmental Assessment Method for different types of buildings (BREEAM);
- BSRIA's *Environmental Code of Practice for Buildings and Their Services*, which has been piloted and is due for open publication in early 1994;
- the CIC Environment Task Group's report, *Our land for our children: An environmental policy for the construction professions*, published in August 1992;
- the CIEC Environment Task Force's report, *Construction and the Environment*, published by the Building Employers' Confederation in May 1992.

In addition, other groups have launched more general, business-related initiatives which seek to secure commitment to continuous improvement in environmental performance by participating companies.

Very many environmental improvements cost nothing to implement except thought, knowledge and commitment, whilst many can also provide cost savings and increase the attractiveness of adopting a positive environmental stance. However, it must also be recognised that some highly desirable environmental improvements will incur additional costs to one or more of the parties involved in a project. The commercial position of the construction industry is rarely if ever conducive to the imposition of such additional costs, whether to clients, designers, contractors or suppliers. As a result, an extra dimension has been added to the environmental challenge – that of improving environmental performance whilst generating new market opportunities and enhancing competitiveness.

The role of designers, contractors and clients

The design professions can and should have a very significant influence on the environmental impact of the projects they design and develop. In broad terms, the research undertaken in preparing these Handbooks indicates that there is adequate information available for a designer who wishes to pursue a policy of minimising the environmental impact of a design: to do so will require commitment and extra effort, but it is not only possible but necessary.

Those responsible for the construction phase must also seek to reduce the impact of their own operations on the environment. Their role is to translate a design into an operational scheme so their ability to redress any inherent environmental weaknesses in a design is limited. In contrast to designers, those responsible for the construction phase can find out *what* they *ought* to do as environmentally responsible contractors but guidance on *how* to achieve it is comparatively lacking. Reference to the guidance in this Handbook will assist in this endeavour.

Clients of the construction industry can also have a marked influence on the environmental performance of their projects. Some clients will not be looking for enhanced environmental performance: the design professions have a particularly important role in persuading such clients towards a more-environmentally responsible approach.

Key issues

From the very wide range of environmental issues facing the industry, a few stand out as very important:

- energy conservation and the need to reduce emissions of so-called greenhouse gases;
- selection of materials against environmental criteria;
- effective conservation of natural resources – designing and constructing with waste minimisation, recycling and salvage in mind;
- pollution control;
- bringing derelict or contaminated sites into beneficial use;
- securing commitment from all concerned to a more environmentally responsible approach;
- the move towards a sustainable environment, an approach which embodies a positive, constructive adoption of the above issues.

Coverage of the Handbook

Readers should note four main points about the coverage of the guidance in this Handbook.

- Many environmental issues affect or are affected by related occupational health and safety concerns. Whilst the impact of the 1974 Health and Safety at Work Act is introduced and its main implications stressed, the Handbook only covers occupational health and safety issues where they have a significant wider environmental impact or may affect the working environment in works or buildings. Mainstream occupational health and safety of construction workers is not covered.

- On the current legal position, the UK situation mainly refers to England and Wales. Generally, legislation which applies to England and Wales also applies to Scotland, with such legislation being modified to take account of the different administrative and judicial systems in Scotland. Where relevant, Scottish primary legislation is listed if it is different from that applying to England and Wales. However, detailed differences in Scottish law are not discussed. No reference is made to Northern Ireland which for the most part has different environmental laws and administrative systems from England and Wales. However, Box 8.9 in *Environmental issues in construction – A review of issues and initiatives relevant to the building, construction and related industries* (CIRIA Special Publications 93 and 94, 1993) gives a listing of the main environmental legislation in England and Wales, Scotland and Northern Ireland.

- A distinction must be drawn between legislative and regulatory requirements and achievable good practice which can, sometimes, go beyond legal minimum standards of environmental performance. The requirements of the former are explicit in the coverage given to relevant regulations. The accepted good practice which is described may also be derived from regulatory requirements. However, in many cases it will be drawn from industry codes of practice, and/or formed from general consensus within the industry and relevant groups, which indicate that higher standards of performance are achievable. Best practice is not defined, since there may be many project-specific criteria which dictate the suitability of the many possible solutions to a problem. It must also be acknowledged that as the regulatory framework is subject to change, so is good practice.

- The Handbook is equally directed at building and civil engineering. Whilst readers will notice that some guidance is specifically related to building design and construction and some to civil engineering schemes, the principles applied are often interchangeable.

Stage C1 Tendering

Stage C1: Tendering provides an introduction to the overall legal position on environmental matters and covers the environmental actions and potential influences of contractors when they are preparing tenders. It is recognised that the primary issues at this stage are commercial ones; the handbook provides guidance on contractors' environmental responsibilities and on areas where they might be able to strike a competitive edge from an environmental standpoint.

C1.1   An introduction to relevant legislation and policy

C1.2   Identification and evaluation of contract environmental requirements

C1.3   Tenderer's potential influence on the environmental aspects of the project

C1.4   Environmental matters to include in a tender

C1.1 An introduction to relevant legislation and policy

Issue C1.1	An introduction to relevant legislation and policy
Current legal position: ▸ C2.1.1 ▸ C2.1.2 ▸ C2.1.3 ▸ C2.1.4 ▸ C2.1.5 ▸ C2.1.6 ▸ C2.1.7 ▸ C2.1.8 ▸ C2.1.9 ▸ C2.1.10 ▸ D3.2.1 ▸ D4.2.2 ▸ D3.7.1	Many of the environmental issues and legal requirements relating to construction projects should be ascertained early in the design and specification phase. There are several areas of environmental legislation and regulations which will have to be considered both at domestic UK law and EC levels. An explanation of the coverage of UK law and in particular the differences between the law in England and Wales, in Scotland and in Northern Ireland is given in the Introduction on page 2. *Legislation* There is a considerable volume of legislation already in existence with the purpose of protecting the environment. There are many areas in which there may be future developments on the law and environmental policy. A distinction can be made between local environmental issues which arise during actual construction such as waste water and solid waste and the wider global environmental context. The legislation mainly takes two forms and these are often interrelated: • the requirement for environmental permits, consents and authorisations e.g. for discharges to water and air, consents governing noise, emissions, hazardous substances and planning permission – breach of these is usually a criminal offence; • the imposition of civil or criminal liability in respect of other matters e.g. the duty of care as respects waste (Section 34 Environmental Protection Act 1990), noise or vibration constituting either statutory nuisance (Sections 79–82 Environmental Protection Act 1990) or common law nuisance. These two elements will affect the inception of the project and its feasibility e.g. whether planning permission is granted or not and, if it is, the conditions attached to it such as the level of clean-up required of any contamination. The following is only a brief summary of the areas of law which should be considered, many of which are dealt with in more detail in specific issue boxes. It should be remembered that no two construction projects are the same. The list below identifies the main areas and it is always necessary to explore the legal issues in each case: • wildlife and countryside law (see C2.1.10); • energy use and efficiency, global warming and climate change (see D3.7.1); • town and country planning law including the environmental assessment of development projects (see D3.2.1); • health and safety law (see C2.1.6); • law on the use of building materials (see D4.2.2) and hazardous substances; • law of waste and recycling (see C2.1.2 and C2.1.3); • pollution control law – emissions to air, water, soil (see C2.1.4, C2.1.5, C2.1.8 and C2.1.9); • public access to information. *Common law* It is also important to consider environmental liability which may arise at common law. This will largely arise where third parties have suffered damage to themselves and/or their property e.g. by contamination of their land or drinking water. The main areas of common law are: • **Nuisance** – the tort of nuisance is basically an act or omission on certain land which unreasonably interferes with or disturbs another person's right of enjoyment of other land. Nuisance may be either public nuisance or private nuisance. Public nuisance is a criminal offence as well as a civil wrong, and only applies if the defendant's act or omission is affecting a significant section of the public as a whole, for example where a contaminated site is polluting a drinking water supply. • **Negligence** – requires proof of fault, i.e. conduct falling below a standard that courts would regard as 'reasonable'. The plaintiff must also show both that he has suffered damage (and not mere economic loss) and that his damage has been caused by the negligence complained of. Additionally, the defendant must owe a duty of care to the particular plaintiff. In the case of a contamination of land, such a duty exists between owners and occupiers of neighbouring property. • **Rule in *Rylands –v– Fletcher*** – imposes strict liability for all damage resulting from a person having brought something on to his land that is not naturally there and which is likely

Issue C1.1	An Introduction to relevant legislation and policy
Current legal position continued:	to 'do mischief' if it escapes. This would include, for example, toxic chemicals, and anything that might cause purely physical damage such as water held back by a dam. It is no defence to prove that all possible precautions have been taken to prevent damage resulting from an escape. • **Trespass** – trespass to land occurs when there has been any unjustifiable intrusion by one person upon the land of another. Contaminants escaping from a contaminated site and entering upon the land of a neighbour would accordingly constitute a trespass.
References to the current legal position:	***The main UK Acts of Parliament are:*** • Alkali Works etc Regulation Act 1906. • Building Act 1984, Building (Scotland) Act 1970, and particularly the 1991 Building Regulations. • Clean Air Act 1993. • Control of Pollution Act 1974. • Environmental Protection Act 1990. • Health and Safety at Work etc Act 1974. • Occupier's Liability Acts 1957 and 1984. • Planning (Hazardous Substances) Act 1990. • Town and Country Planning Act 1990, Town and Country Planning (Scotland) Act 1972, Planning and Compensation Act 1991, and Planning (Consequential Provisions) Act 1990. • Water Resources Act 1991. • Water Industry Act 1991. ***Other references:*** • The numerous regulations under the above Acts of Parliament. • Environmental Information Regulations 1992 (SI 1992 No. 3240). • Uff, J., *Construction Law*, 5th Edition, Sweet & Maxwell, 1991.
Policy and forthcoming legislation:	EC policy is broadly contained in the 5th Environmental Action Programme entitled *Towards Sustainability*. The programme is a policy document for environmental legislation to the year 2000. There are five target areas tourism, industry, energy, transport and agriculture. Current EC policy is a result of earlier Action Programmes. Policy documents may also focus on a specific area for example the EC has policies on waste and energy in addition to the 5th Environmental Action Programme. Current UK policy is contained in the White Paper *This Common Inheritance* and the yearly reports on implementation of the proposals contained in that document. Policy focuses include aspects of climatic change and global warming, economic instruments and the introduction of market mechanisms to promote environmental protection, and energy efficiency. A paving bill is expected in the 1993/94 Session which will contain provisions for the establishment of an Environment Agency for England and Wales and a Scottish equivalent. The proposed Construction (Design and Management) Regulations – CONDAM – will apply to all projects that include construction work. The regulations set out responsibilities for developers, designers and contractors in respect of health and safety during construction.
Policy references:	**EC** • Fifth Action Programme on the Environment, *Towards Sustainability* (OJ C138 17/5/93). • Draft Directive on Landfill of Waste (Com (91) 102). • Green Paper, *Proposals for remedying environmental damage*, 1993. • Draft Proposals for a Directive on a tax on carbon dioxide emissions, energy efficiency and the promotion of renewable energy (COMs (92) 226, 182 and 180). **UK** • Government White Paper: *This Common Inheritance – Britain's Environmental Strategy*, 1990 and *Yearly Reports* 1991 and 1992. • Health & Safety Commission, *Proposals for Construction (Design and Management) Regulations and Approved Code of Practice*, Health and Safety Executive, 1992.

C1.2 Identification and evaluation of contract environmental requirements

Issue C1.2	Identification and evaluation of contract environmental requirements
Background:	As the environmental considerations in new building and civil engineering projects, both explicit and implied, become more important, increasing use of special terms covering environmental issues in contract documents can be expected. At present, most contract documents only include explicit environmental issues as part of the specification, for example in making specific provisions for the exclusion of CFCs or particular solvents, in specifying particular 'environmentally-friendly' materials or ways in which particular legal requirements are to be met. However, general environmental provisions, such as a total ban on the use of CFCs in a project, or the use only of diesel-engined vehicles, continue to be slowly introduced, although few in number at present.
	Such provisions are anticipated to become commonplace as environmental management systems (EMS) based on BS7750 are developed and clients come to expect contractors to have such systems in place. This is forecast by some to mirror the development of quality management systems based on BS5750, with some clients requiring any contractor wishing to tender to have established a certificated quality management system or a total quality management system to BS7850. However, since BS7750 has only recently been piloted, and a revised edition is awaited, even the most environmentally conscious contractors cannot yet have a BS7750-derived EMS in place.
	Separately, as the later sections of this document demonstrate, the legal and regulatory environmental requirements on contractors and others involved on site are increasing. Clients and their professional advisors should be explaining to prospective contractors the main environmental facets of the project and the legal requirements contractors will be expected to meet. In addition, if a formal environmental policy for the project has been agreed, or formal statements of specific environmental requirements drawn up, they should be included in tender documents and a formal requirement placed on tenderers to make specific provision for meeting the policy requirements.
Background references:	• Miller, S., *Going Green*, JT Design Build, Bristol. • *Environmental issues in construction – A review of issues and initiatives relevant to the building, construction and related industries*, CIRIA Special Publications 93 and 94, 1993, Chapter 8, Legislation and policy. • CIC Environment Task Group, *Our land for our children: an environmental policy for the construction professions*, Construction Industry Council, August 1992. • CIEC Environment Task Force, *Construction and the Environment*, BEC, May 1992. • BS7750: 1992 *Specification for Environmental Management Systems*, BSI. • BS5750: 1987 *Quality Systems*, BSI, Milton Keynes. • BS7850: 1992 *Total Quality Management*, BSI, Milton Keynes.
Good Practice: ► C2.2.1 ► D5	*When reviewing the tender documents from an environmental standpoint, look out for, and consider the implications for your tender of:* • any project environmental policy or other statement of environmental requirements; • any requirement to appoint an individual to have specific responsibilities for environmental matters on the project (often referred to as an Environmental Manager); • any bans included in the specification on processes, procedures and/or materials, eg a ban on CFCs, HCFCs, the use of non-diesel vehicles; • look carefully at the specification for particular environmental requirements and consider how your purchasing policies will cope with them; • look for, or evaluate the likely need for, a dedicated (part- or full-time) environmental manager for the project.
Good practice references and further reading:	• No references have been identified as giving specific guidance on how to include or evaluate environmental matters in construction tender documents. However, the following guide gives cogent guidance on inclusion of health and safety requirements which will in time be extendable to environmental matters. • European Construction Institute, *Total project management of construction safety, health and environment*, Thomas Telford, 1992. • BS7750 as in background references.

C1.3   Tenderer's potential influence on the environmental aspects of the project

Issue C1.3	Tenderer's potential influence on the environmental aspects of the project
Background: ► C1.2	In addition to the issues raised in C1.2, contractors are increasingly seeking to use environmental issues and/or an environmental stance as a marketing advantage. Some are adding an environmental review of the project to their tender submission, with appropriate additional costs (and occasionally cost reductions) given for each potential environmental improvement identified. Although the environmental policy for a project (explicitly stated or not) will have been set by the client and professional advisor, contractors can have a positive environmental influence on a project by, for example, suggesting alternative materials which the contractor believes to be more environmentally acceptable than those specified. Equally, an alternative construction method to that implied by the tender documents may be less disruptive to neighbours, consume less of a scarce resource or provide other environmental benefits. Increasingly, despite the present difficult economic climate, more clients are prepared to accept some modest premium for an acknowledged environmentally benign project, so there are opportunities with such clients for well-prepared contractors to improve their chances of success by adding to their tender some environmental commentary and /or proposals to their submission.
Background references:	• CIC Environment Task Group, *Our land for our children: an environmental policy for the construction professions*, Construction Industry Council, August 1992. • CIEC Environment Task Force, *Construction and the Environment*, Building Employers' Confederation, May 1992. • Construction Industry Environmental Forum, *Environmental management in the construction industry*, Notes of meeting held on 22/09/92, CIRIA, 1992. • Miller, S., *Going Green*, JT Design Build, Bristol.
Good Practice:	***When reviewing the tender documents:*** • consider the extent to which you consider it likely that the client and/or professional advisor will welcome your taking an environmentally positive stance in your tender; • consider the environmental policy for the project (explicitly stated or not), consider how you will be able to meet its requirements, and decide whether to include an environmental addendum to your submission; • use available information on environmentally less-damaging products and materials; • review the technical specification and identify areas where you may be able to offer a more environmentally acceptable alternative; • indicate the steps you propose to take to minimise the environmental risks to the client and site neighbours; • if there is no stated environmental policy for the project, consider including your own company policy as part of, or as an extra to, the submission.
Good practice references and further reading:	• As good practice references, plus: • Hall, K., and Warm, P., *Greener Building Products & Services Directory*, Association for Environment Conscious Building Directory, Second Edition, 1993. • BS7750: 1992 *Specification for Environmental Management Systems*, British Standards Institution, Milton Keynes. • Construction Industry Environmental Forum, *Green Clients*, Notes of Workshop held on 30/11/92, CIRIA, 1992. • European Construction Institute, *Total project management of construction safety, health and environment*, Thomas Telford, 1992. • Ove Arup & Partners, *The Green Construction Handbook – A Manual for Clients and Construction Professionals*, JT Design Build, Bristol, 1993

C1.4   Environmental matters to include in a tender

Issue C1.4	Environmental matters to include in a tender
Background:	See C1.2 and C1.3.
Background references:	See C1.2 and C1.3.
Good Practice:	*Having undertaken the suggested actions in C1.2 and C1.3:* • ensure that you have explicitly provided for how you will meet the environmental requirements included in the tender documents; • include your own company's environmental policy and highlight any environmental purchasing policy; • if you have an environmental management system (EMS) in place, include it in your submission and demonstrate how it would be applied to the project and the benefits that would accrue; • if you have no EMS but have decided to appoint an environmental manager, ensure that their identity, intended role, the planned relationships on environmental matters with the Agent and Resident Engineer, and their relevant experience are all clearly presented; • if you are proposing alternative materials on environmental grounds, ensure that sufficient justification for the proposed change is clearly presented or referenced; • if you have decided to include an addendum of environmental proposals, decide the basis on which they are offered and then present the financial implications in a manner best suited to your overall tender; • if you feel some of the environmental requirements are unnecessarily onerous, or perhaps not achievable, show the impact of these conditions but in a positive way and, only as a last resort, ensure your bid is suitably qualified.
Good practice references and further reading:	• No publicly-available references have been identified as giving guidance on how to include specific environmental matters in a tender submission for a building or civil engineering project. • However, Croner's *Environmental Management*, with quarterly amendment service, (Croner Publications Ltd, First Edn, October 1991) provides ample guidance on the process of becoming an environmentally responsible organisation irrespective of the nature of the business, and may prove very useful to estimators and project managers, in conjunction with this handbook, in identifying their responsibilities and the opportunities in this area.

Stage C2 Project planning and contract letting

Stage C2: Project planning and contract letting covers mainly from a contractor's point of view, the environmental issues to be faced when negotiating a contract and planning, before going on site, the operation of a new project. Section C2.1 provides introductions to the legal position on ten important areas.

C2.1 Legislation and policy

C2.2 Overall purchasing policies

C2.3 Green management of a site

C2.4 Pollution control strategies

C2.5 Contaminated land

C2.6 Relations with relevant bodies and groups

C2.7 Special considerations for design and build projects

C2.1 Legislation and policy

Issue C2.1.1	The Environmental Protection Act 1990
Current legal position: ▶ C2.1.2 ▶ C2.1.3 ▶ C2.1.5 ▶ C2.1.8 ▶ C3.5.4 ▶ C2.1.9	The Environmental Protection Act 1990 (EPA) was enacted on 1 November 1990 and most of it is now in force. The Act coincided with growth of public and media interest and concern for the environment at local, national and global levels. It is also important to recognise the role of EC legislation in shaping many parts of the EPA. It should really be considered as a framework to which the many subsequent regulations have added the detail and the administrative mechanisms required for pollution control. The EPA consolidated and amended the previous piecemeal pollution controls and will inevitably impinge upon many aspects of the construction phase of building and civil engineering projects. The following issues are dealt with specifically in other issues: • the control of waste on land – see C2.1.2; • the responsibility for the waste produced (the 'Duty of Care') – see C2.1.3; • liability in respect of pollution and contaminated land – see C2.1.8; • the control of noise emitted from a construction site – see C2.1.5 and C3.5.4; • air pollution control – see C2.1.9. The EPA also introduces other general provisions which are important, in particular the following. • **Liability of directors and officers of companies, and of parent companies for their subsidiaries** – This liability may arise where any offence under the EPA has been committed by a company with the consent or connivance of any director or officer or is attributable to any neglect by such a person; then he will also be guilty of the offence. Further, a parent company involved in the management of its subsidiary will be treated as though it were a director of that subsidiary and liable for offence. • **Statutory nuisances** – The EPA consolidated all statutory legislation on nuisances and provided a more streamlined procedure for dealing with them. Categories of statutory nuisance are wide-ranging and include smoke, gases or fumes which are emitted; dust, steam or smell arising from an industrial trade or business premises; any accumulation or deposit emitted. Any of these matters may be a nuisance if it is prejudicial to public health. A local authority may raise an abatement notice or an aggrieved person may seek an order requiring abatement from the Magistrates Court. • **Dangerous substances** – The Secretary of State is given power under the EPA to make regulations to prohibit or restrict the importation, use, supply or storage of injurious substances or articles. • **Access to information** – Several provisions of the EPA require registers to be kept and information to be available to the public. The majority of the offences under the EPA are subject to a maximum fine of £20,000 in the Magistrates' Court. More serious cases which are referred to the Crown Court are subject to unlimited fines and/or imprisonment for up to 2 years. Previously offences under environmental legislation were subject to a maximum fine of £2,000. It is important to mention the increased emphasis on enforcement. It is clear that enforcement agencies such as the NRA or Her Majesty's Inspectorate of Pollution are becoming less tolerant of industry failure to act on events which may threaten human health or the environment. In general the EPA does not apply to Northern Ireland.
References to the current legal position:	• Environmental Protection Act 1990. • Environmental Protection Act 1990, Commencement Orders Nos.1–13 (1990–1993).

Issue C2.1.1	The Environmental Protection Act 1990
Policy and forthcoming legislation:	UK environmental policy, both existing policies and potential future policy and strategy, are explained in the White Paper entitled *This Common Inheritance – Britain's Environmental Strategy*. The White Paper commits the Government to improving the environment in many ways and indications of possible future policy and legislative initiatives are included.
	The White Paper contains over two hundred proposals for action and progress towards these is published in the yearly reports. Of most importance in this context are encouragement of the prudent and efficient use of energy and other resources, the aim to make sure that air and water are clean and that controls over waste are maintained and strengthened, and that noise pollution control is strengthened. Guidance on the various aspects of the EPA is given in DoE circulars issued on specific subjects from time to time.
Policy references:	• Government White Paper: *This Common Inheritance – Britain's Environmental Strategy*, 1990. • Government White Paper: *This Common Inheritance – Britain's Environmental Strategy – Second Year Report*, 1992.

Issue C2.1.2	Waste legislation and policy

Current legal position:	Proper waste management to avoid legal liability is essential to the construction industry. Part II of the Environmental Protection Act 1990 (EPA) and related secondary legislation strengthens and restructures the regulation of waste collection and disposal, the licensing of disposal and other operations involving controlled waste. The EPA replaces the system of waste disposal licences under the Control of Pollution Act 1974 (COPA). However, currently, waste management is partly regulated by COPA and partly regulated by the EPA depending on the particular provision. The implementation of the waste management licensing provisions under the EPA has been delayed but they are expected to be in place by May 1994. The EPA has created new waste regulation authorities which are responsible for the regulation of waste disposal, while local authority waste disposal operations have been transferred into arms length companies called local authority waste disposal companies (usually called LAWDCS).

What is 'waste'?

The first point to consider is whether material is 'waste' or not. Once this has been established, it is then necessary to consider whether the material is 'controlled waste' to which many of the legislative provisions apply. Both the terms 'waste' and 'controlled waste' are defined in the legislation and have also been the subject of interpretation by the courts.

The definition of waste differs between EC legislation and UK legislation. The EC Framework Directive on waste (75/442/EEC as amended) defines waste as 'any substance or object in the categories set out in Annex I (which lists waste categories e.g. products for which the holder has no further use, contaminated materials, off specification products) which the holder discards or intends or is required to discard'.

Section 75 of the EPA defines waste as 'any substance which constitutes a scrap metal or an effluent or other unwanted surplus substances arising from the application of any process; and any substance or article which requires to be disposed of, has been broken, worn out, contaminated or otherwise spoilt'.

The interpretation of the definition of waste by the courts has consistently shown that provided material is regarded by the producer as waste or is surplus to requirements it is irrelevant that the same material has value to another: it is still waste. Therefore, soil which is excavated or obtained for the purposes of construction and for use as fill or re-use could, depending on the circumstances, be waste in the eyes of the producer of that material even though in the eyes of the person using the material it is a valuable resource. However, material which is brought onto a site, and which is then not used for the purposes for which it was brought on (i.e. surplus fill) could become waste at the time it becomes surplus to requirements. What is crucial is the view of the waste regulation authority who have a discretion to interpret the legislation as they see fit. Unfortunately, it is not possible, because of confusion about the definition of waste and the difference in interpretation between waste regulation authorities in different areas, to make any general rules relating to the construction industry as to what is and what is not waste. Therefore, the circumstances in each case must be considered.

'Controlled waste' is defined for the purposes of both the COPA and the EPA as household, industrial or commercial waste. Regulations have been made under both of these acts to give detail to the three categories. Generally speaking waste from works of construction or demolition (including preparatory work) and waste from tunnelling or excavation (which are not defined expressions) are specified as industrial waste and are therefore controlled waste.

Site licensing

Currently, a site licence under COPA is required for the disposal of controlled waste. However, the EPA imposes a new waste management licensing scheme (not yet in force) the detail of which is yet to be decided. The requirements of the EPA are likely to have a greater impact on construction sites and may require a waste management licence (WML) for activities which currently do not need to be licensed under the COPA. Because the Waste Management Licensing Regulations have not yet been agreed, it is difficult to say how these regulations will apply and affect construction sites.

Issue C2.1.2	Waste legislation and policy continued

Current legal position continued:	Section 33 of the EPA requires a person to hold a WML if he is to treat, keep or dispose of controlled waste. This provision is wider than the previous COPA provision, whereby a licence was only required for the final deposit of waste. An applicant for a WML will need to demonstrate that he is a fit and proper person who is technically and financially competent. WMLs are granted by a waste regulation authority on such terms and subject to such conditions as they think fit.

Section 33 creates several criminal offences. These include: the depositing of controlled waste or knowingly causing or knowingly permitting controlled waste to be deposited, in or on any land, unless a WML is in force and the deposit is in accordance with that licence; the treating, keeping or disposing of controlled waste or knowingly causing or knowingly permitting controlled waste to be treated, kept or disposed of, in or on any land (or by means of any mobile plant), except in accordance with the WML; and the treating, keeping or disposing of controlled waste in a manner likely to cause pollution of the environment or harm to human health.

Licensing exemptions

Under the proposed Waste Management Licensing Regulations there is an exemption from two of the above offences where construction waste is used in the manufacture of certain construction products such as bricks, plaster board or aggregate.

► C2.1.3

There is also a specific exemption from the requirement to have a WML in the first place relating to the keeping or deposit of waste from demolition or construction work. The exemption applies to:

- waste which is to be used for construction work to be carried out on the land in question; and
- the waste is not kept on the land for more than three months prior to the commencement of construction work; and
- the total quantity of waste used is not 'excessive'.

In these circumstances a WML would not be required.

However, construction operations may be subject to the requirement to obtain a WML especially in relation to the keeping (as opposed to disposal) of waste where construction waste is merely accumulated or stored on site and does not fall into the above exemption, for example if waste is kept on site before removal for disposal or waste is kept on site for more than three months. The keeping of waste is deemed by the DoE to cover a range of situations, for example where it is retained in the possession or control of the holder, including waste storage.

Civil liability

Independent of the requirement for a waste management licence a developer may incur civil liability for pollution to the environment from controlled waste resulting in damage or injury to people and their property.

Duty of care

There is a statutory duty of care as respects waste (Section 34), to take all reasonable steps to prevent the commission of an offence under Section 33 of the EPA. The aim of the duty of care is to increase the responsibility of persons in respect of waste they hold unless it is transferred to authorised parties and disposed of safely. Breach of the duty of care is a criminal offence.

Special waste

For certain dangerous or intractable waste ('special waste'), the Controlled Waste (Special Waste) Regulations 1980 (SI 1980 No.1709) provide for a system of consignment notes, as well as for registers and site records. These regulations provide a list of substances that are designated as special waste if they are dangerous to life (e.g. asbestos and lead compounds) or have a flash point of 21 degrees centigrade or less. Failure to comply with these regulations on the part of a producer, disposer or carrier of special waste is an offence liable on summary conviction to a fine not exceeding £5,000 or on a conviction on indictment to an unlimited fine and to imprisonment for a term not exceeding two years.

Continued overleaf

Issue C2.1.2	Waste legislation and policy continued

Current legal position continued:	*Transport of waste* Under the Control of Pollution (Amendment) Act 1989 it is an offence for a person to transport demolition or construction waste to or from any place in Great Britain with a view to making a profit, in the course of his business, without being registered. For transporting waste between different places within the same premises registration is not necessary. Regulations made under this Act govern registration and enforcement. Application for registration is made to the relevant waste regulation authority. An application cannot be made by a person who has been convicted of one of the offences listed in the regulations. The fine for non-registration is up to £5000, after which a subsequent application could be refused. Depending on circumstances construction enterprises may need to consider registering with the waste regulation authority. *Trans-frontier shipment of waste* There are also regulations relating to the movement of hazardous waste into and out of the UK.
References to the current legal position:	**EC** • Framework Directive on Waste 75/442/EEC (as amended by 91/156/EEC). • Directive on Hazardous Waste 91/689/EEC. • Directive on the Trans-frontier Shipment within the EC of hazardous and toxic waste 84/631/EEC. • Regulation on the supervision and control of shipments of waste into and out of the EC (259/93) (to apply member states from May 1994). **UK** • Control of Pollution (Amendment) Act 1989. • Environmental Protection Act 1990. • Control of Pollution (Special Waste) Regulations 1980 (SI 1980 No.1709) (as amended by SI 1988 No.1790). • Trans-frontier Shipment of Hazardous Waste Regulations 1988 (SI 1988 No.1562). • Collection and Disposal of Waste Regulations 1988 (SI 1988 No.819). • Disposal of Controlled Waste (Exceptions) Regulations 1991 (SI 1991 No.508). • Controlled Waste (Registration of Carriers and Seizure of Vehicles) Regulations 1991 (SI 1991 No.1654). • Controlled Waste Regulations 1992 (SI 1992 No.588) (as amended by SI 1993 No.566). • Environmental Protection (Waste Recycling Payments) Regulations 1992 (SI 1992 No.462) (as amended by SI 1993 No.445). • The draft Waste Management Licensing Regulations, 1992.
Policy and forthcoming legislation:	It is generally accepted that the production of waste is an inevitable by-product of industrial development. However, Government policy is aimed at waste prevention and minimisation to reduce harm to the environment. This includes emphasis on recycling where possible, with the aid of financial incentives. Government policy can be found in DoE Circulars; Waste Management Papers; local authority waste disposal and management plans and related Planning Policy Guidance Notes; local authority waste recycling plans. The regulations relating to special waste are intended to be revised when the EC has drawn up a list of 'hazardous wastes'. There is also a proposed EC Directive on civil (strict) liability for damage caused by waste to people, property and to the environment.

Issue C2.1.2	Waste legislation and policy continued
Policy references:	• EC Green Paper: *Remedying Environmental Damage* (March 1993). • Government White Paper: *This Common Inheritance – Britain's Environmental Strategy*, 1990. • Government White Paper: *This Common Inheritance – Britain's Environmental Strategy*, Second Year Report, 1992. • DoE Circular 13/88, The Collection and Disposal of Waste Regulations. • DoE Circular 16/89, The Trans-frontier Shipment of Hazardous Waste Regulations 1988: the Control of Pollution (Special Waste) (Amendment) Regulations 1988. • DoE Circular 8/91, Competition for local authority waste disposal contracts and new arrangements for disposal operations. • DoE Circular 10/91, Separation of local authority waste regulatory and waste disposal functions. • DoE Circular 11/91, Controlled Waste (Registration of Carriers and Seizure of Vehicles) Regulations 1991. • DoE Circular 4/92, The Waste Recycling Payment Regulations 1992. • DoE Circular 14/92, The Environmental Protection Act 1990, Part II and IV: the Controlled Waste Regulations 1992. • DoE Waste Management Paper No.1, A Review of Options, 1992. • DoE Waste Management Paper No.4, The Licensing of Waste Facilities, 1988 2nd Edition (in process of being revised). • DoE Waste Management Paper No.23 Special Waste: A Technical Memorandum Providing Guidance on their Definitions, 1983. • DoE Waste Management Paper No.28 Recycling: A Memorandum Providing Guidance to Local Authorities on Recycling, 1991. • There are several DoE Waste Management Papers dealing with specific wastes such as solvent waste (No.14); wood preserving waste (No.16); asbestos waste (No.18).

Issue C2.1.3	Duty of care for waste

| Current legal position:

▸ C2.1.2 | The 'duty of care as respects waste' was introduced on 1 April 1992 under Section 34 of Part II of the Environmental Protection Act 1990 (EPA). The aim of the duty of care is to make fly tipping more difficult by ensuring the safe management of controlled waste, from its generation to its final disposal.

Who does the duty of care apply to?

The duty of care applies to all persons who 'import, produce, carry, keep, treat or dispose of waste, or as brokers have control of such waste'. The duty of care therefore applies to a wide range of people who may be regarded as holders of controlled waste irrespective of size of business or type of waste produced. It will therefore apply to the construction industry. Controlled waste is defined under the EPA as 'household, industrial and commercial waste'. The Controlled Waste Regulations (SI 1992 No. 588) provide a more detailed definition of the three main categories of waste for the purposes, *inter alia*, of the duty of care provisions.

In the construction and demolition industries where there is often sub-contracting there may be uncertainty as to who is the 'producer' of waste. It is the view of the DoE that the producer of waste may be regarded as the person undertaking the works that give rise to waste, not the person who issues instructions or lets contracts which give rise to waste. Guidance on this is in the DoE Circular 19/91 paragraphs 17 and 18. In practice it is likely that each contractor will either be producing or carrying waste away and will be subject to the duty of care.

However, the duty of care also applies to 'brokers' of waste. These are people who are not holders of waste but who arrange transport of waste and as such have control over what happens to it. Hence, a developer who arranges directly with a haulier for transport of waste may be exposed to the duty. Liability may extend to company directors and to any parent company.

Requirements of the duty of care

The duty of care requires that the holders of waste take all reasonable steps to:

• prevent the unauthorised or harmful deposit, treatment, keeping or disposal of waste without a waste management licence or in contravention of a condition of a licence or in a manner likely to cause pollution of the environment or harm to human health by any other person;
• prevent the escape of waste from their own control or that of another person, that is to contain waste;
• ensure that waste is only transferred to an authorised person or to a person for authorised transport purposes;
• ensure that the transferred waste is accompanied by an adequate written description.

These obligations are limited by the requirement only to take measures which are 'reasonable in the circumstances' and applicable to the holder of the waste. The circumstances which will determine what is reasonable are:

• what the waste is;
• the dangers it presents;
• how it is dealt with; and
• what the holder might reasonably be expected to know or foresee.

The capacity of the holder is determined by who he is, how much control he has over what happens to waste and what his connection with the waste is. |

Issue C2.1.3	Duty of care for waste
Current legal position continued	The Duty of Care Regulations require both transferors and transferees of waste to keep waste records (for at least 2 years). This should be done by a transfer note detailing the waste (description, quantity, type), the container, name/address and capacity of person transferring the waste (eg. producer, carrier etc.) and capacity of person receiving waste, date and place of transfer and the waste regulatory authority. The transfer note should be signed by both the transferee and the transferor. Special waste is also subject to the duty of care. Furthermore, compliance with the duty of care does not mean that the special waste regulations should not be complied with. An 'authorised person' includes a waste collection authority, a person who holds a waste management licence, or any person registered as a carrier of controlled waste under Section 2 of the Control of Pollution (Amendment) Act 1989. *Penalty* Failure to comply with the duty of care is a criminal offence. The offence is punishable by a fine of up to the statutory maximum (£5,000) on conviction in a Magistrates' Court and an unlimited fine on conviction in a Crown Court.
References to the current legal position:	• Environmental Protection Act 1990, Sections 33 and 34. • Environmental Protection (Duty of Care) Regulations 1991 (SI 1991 No. 2839). • Controlled Waste Regulations 1992 (SI 1992 No. 588) (as amended by SI 1993 No. 566). • Control of Pollution (Amendment) Act 1989. • Controlled Waste (Registration of Carriers and Seizure of Vehicles) Regulations 1991 (SI 1991 No. 1624).
Policy and forthcoming legislation:	The general aim of government policy in respect of waste is waste minimisation. However, the duty of care is aimed at ensuring that anyone who produces or handles waste cannot easily escape responsibility for what happens to it, thus embodying the 'polluter pays' principle. A Code of Practice, which has statutory standing (and may be admitted as evidence in Court), sets out practical guidance to holders of waste subject to the duty of care on how to comply with the duty. In practice it is likely that all contractors working on a site will be subject to the duty and the code advises that reasonable steps should be taken to ensure that all contractors employed comply with the duty of care. The Code of Practice sets out step-by-step advice on complying with the duty of care. This includes identifying and describing the waste, keeping the waste safely, transferring waste to the right person, receiving waste, keeping records, and checking up and seeking expert help and advice.
Policy references:	• Government White Paper: *This Common Inheritance – Britain's Environmental Strategy*, 1990. • Government White Paper: *This Common Inheritance – Britain's Environmental Strategy*, Second Year Report, 1992. • DoE, *Waste Management: The Duty of Care – A Code of Practice*, 1991. • DoE Circular 19/91, The Environmental Protection Act 1990 – The Duty of Care • DoE Circular 14/92, The Environmental Protection Act 1990 – Parts II and IV – The Controlled Waste Regulations, 1992.

Issue C2.1.4	Water legislation and policy
Current legal position:	Water and drainage law has an impact on the construction industry. The two main statutes to consider are the Water Resources Act 1991 (WRA) and the Water Industry Act 1991 (WIA). These acts consolidated the environmental legislation relating to water, although the substance of the law has generally not changed. *The National Rivers Authority (NRA)* – The NRA will be consulted by the local planning authority in circumstances where the proposed development is likely to have an impact on water resources or drainage. This includes major developments, development in the vicinity of main river watercourses, development in the floodplain or works which may affect the bed, banks or flow of any watercourse. A Land Drainage Consent may be required. The NRA also has statutory duties in relation to the conservation and enhancement of the aquatic environment. *Water Resources Act 1991* • **Discharge Consent** – The WRA has a system of discharge consents for discharges into controlled waters. The NRA is the body which grants discharge consents under the WRA. The NRA will consult with local authorities and statutory water companies in the area prior to granting consent. The consent may be unconditional or with conditions. To discharge without a consent or in breach of a consent is an offence. • **Abstraction Licence** – The NRA also authorise the abstraction of water from a source of supply and conditions may be attached to an abstraction licence. To abstract water without a licence or in contravention of its terms is an offence. • **Impounding Licence** – Where works are proposed to create a reservoir or artificial lake which requires impounding water behind a dam or weir, or where works are to divert the flow of inland water an 'impounding licence' is required. Applications are made to the NRA. Failure to have or comply with an impounding licence is an offence. • **Water Pollution Offence** – Under the WRA, it is an offence to cause or knowingly permit any matter which is itself polluting or has a polluting effect to enter either watercourses or groundwater. Because of this strict liability offence site safety and security is important to prevent the escape of pollutants. It is a defence to the offence above if the discharge was authorised by a valid discharge licence or if the discharge was necessary to avoid harm to human life or health and steps were taken to minimise the effects of the discharge. On conviction in the Magistrates Court the fine may be up to £20,000 and/or up to 3 months imprisonment. In the Crown Court the fine is unlimited and/or imprisonment for up to 2 years. • **Clean-up** – Under the WRA, the NRA has the power to undertake work to prevent or remedy pollution. These powers are mostly used in emergency situations. If the NRA conducts a clean-up, it can charge the person who caused or knowingly permitted the situation, for the reasonable costs of clean-up operations. *Water Industry Act 1991* – Discharges of trade effluent to the public sewer system are governed by the WIA. Trade effluent is any liquid (with or without suspended particles) which is wholly or partly produced in the course of trade or industry carried out at premises used or intended to be used for carrying on any trade or industry. A consent or agreement to discharge must be made with the local water or sewerage company. Under the WIA sewerage undertakers have a general duty to allow the discharge of trade effluent. However, in practice a discharge can still only occur with either statutory consent or an agreement. A consent or agreement may be unconditional or granted subject to wide ranging conditions. Breach of conditions is a criminal offence. If the trade effluent contains certain hazardous substances, known as 'red list' substances, HMIP will determine whether or not to grant consent as the general duty to accept the discharge of trade effluent does not apply. *Salmon and Freshwater Fisheries Act 1975* – It is an offence to cause or knowingly permit any liquid or solid matter to enter into waters (or their tributaries) containing fish to such an extent that waters are injurious to fish, spawning grounds, spawn or food of fish (similar to WRA offence). *Byelaws* – The various water statues enable statutory water undertakers to make byelaws in respect of watercourses which belong to them, to protect against pollution. It is always advisable to check with the local statutory water undertaker whether any byelaws exist.

Issue C2.1.4	Water legislation and policy

References to the current legal position:	**EC** • Directive on Pollution caused by the Discharge of Certain Dangerous Substances into the Aquatic Environment – 76/464/EEC (Several 'daughter' directives for specific substances). • Directive on Ground Water 80/68/EEC. **UK** • Water Resources Act 1991. • Water Industry Act 1991. • Control of Pollution Act 1974. • Salmon and Freshwater Fisheries Act 1975. • Rivers Prevention and Pollution Acts 1951 and 1961. • Water (Scotland) Act 1980. • Rivers (Prevention of Pollution) (Scotland) Acts 1951 and 1956. • Sewerage (Scotland) Act 1958. • Trade Effluents (Prescribed Processes and Substances) Regulations 1989 (SI 1989 No. 1156) (as amended by SI 1990 No. 1629).

Policy and forthcoming legislation:	Government policy is aimed at improving the quality of UK drinking water, rivers and surrounding seas. These objectives have been in four main ways:
	• by establishing organisations to:
	  – supply water – the Water plcs
	  – manage water resources – the NRA
	  – monitor and control quality – the DWI;
	• by setting standards for water quality;
	• by regulating to prevent pollution;
	• by establishing guidelines for recreation and wildlife conservation.
	
	The NRA has produced a groundwater protection policy. It contains eight policy statements which aim to ensure that risks to groundwater are dealt with and that the approaches of the ten NRA regions are compatible. Of particular relevance are the policy statements on waste disposal on land and waste disposal effluents.
	
	To prevent pollution of water courses the NRA and HMIP have called for better fire containment facilities for the prevention of pollution of water courses by waste water run off from fire sites.
	
	Under the WRA, the NRA can draw up regulations covering a wide range of issues and has already issued many on pollution prevention.

Policy references:	• Government White Paper: *This Common Inheritance – Britain's Environmental Strategy*, 1990. • Government White Paper: *This Common Inheritance – Britain's Environmental Strategy*, Second Year Report, 1992. • DoE Circular 17/84, Water and the Environment – The Implementation of Part II of the Control of Pollution Act 1974. • NRA, Corporate Plans (published every year). • NRA, *Policy and Practice for the Protection of Groundwater*, 1993. • NRA, *Annual Reports* (every year). • DoE Circular 20/90, EC Directive on the protection of ground water against pollution caused by certain dangerous substances. • NRA, *Guidance notes for local planning authorities on the methods of protecting the water environment through development plans*, Draft May 1993, NRA Bristol.

Issue C2.1.5	Noise legislation and policy
Current legal position:	Noise pollution from a construction site may affect many people from employees and contractors on the site to people living and working in the locality. Consequently legislation reflects these two elements and addresses noise control for the benefit of workers and third parties. Legal powers to control noise derive from the EC, UK common law and from UK statute. • *Control of Pollution Act 1974 (COPA)* contains specific requirements to control noise from construction sites. This includes the erection, construction, alteration, repair or maintenance of buildings, structures or roads and demolition works. COPA deals with noise abatement and reduction on construction sites. Local authorities can give a notice specifying hours of work, plant or machinery to be used and the level of noise to be emitted where work is being or is going to be conducted on a construction site. Alternatively an application may be made to the local authority for prior consent for works on construction sites. Local authorities may also designate noise abatement zones. In such a zone, noise levels must not be exceeded, and the local authority may issue a notice requesting a reduction in noise levels or consent to noise exceeding registered levels. • *Environmental Protection Act 1990 (EPA)* defines noise (including vibration) emanating from premises, which is prejudicial to health, as a statutory nuisance. There is no level of noise which is a statutory nuisance – it will always depend on the facts of the case. The local authority has the power to serve notice requiring abatement of the statutory nuisance and prohibiting or restricting its recurrence. The notice may also specify remedial work to achieve these aims. A third party may also seek an order from the Magistrates' Court requiring abatement, the prohibition of recurrence and the execution of remedial works in respect of the nuisance. Breach of an abatement notice or abatement order is a criminal offence. Noise nuisance from industrial, trade or business premises may be subject to a fine of up to £20,000. In Scotland the EPA does not apply – but COPA imposes similar provisions. • *Common law private or public nuisance* provides a possible remedy for noise pollution. Nuisance is the unlawful interference with a person's use or enjoyment of land, or of some right over, or in connection with it. A third party may commence a civil action against a construction company for noise which constitutes a nuisance. An injunction may be awarded and/or damages given. • *Workplace noise* is governed by health and safety legislation. The Noise at Work Regulations 1989 were enacted under the Health and Safety at Work etc. Act 1974 to implement EC Directive 86/188/EEC. These regulations protect workers from the risks related to exposure to noise at work. They place responsibility on the employer to ensure adequate protection and to reduce the risk to hearing by minimising the risk of damage and reducing exposure to noise e.g. by setting action levels, defining ear protection zones, the provision of ear protectors. • There are also regulations which *inter alia* control noise from road vehicles such as lorries. The level of noise which various types of vehicle can make when used on the road are limited. Local authorities may also limit numbers and the size of vehicles on roads. The routing of heavy lorries may be controlled by local traffic authorities under the Road Traffic Regulation Act 1984. • *Planning and noise* – Noise is often an issue which is addressed in a planning condition to prevent or minimise noise disturbance, for example by setting noise limits or hours of work. Some planning authorities may require applications for particularly noisy developments to be publicised. Breach of a condition may lead to enforcement proceedings and the possible commission of an offence under the town and country planning legislation.

Issue C2.1.5	Noise legislation and policy

References to the current legal position:	**EC** • Directive on the Protection of Workers from the Risks Relating to Exposure to Noise at Work (86/188/EEC). **UK** • Control of Pollution Act 1974. • Environmental Protection Act 1990. • Road Traffic Act 1974. • Road Traffic Regulation Act 1984. • Health and Safety at Work etc Act 1974. • Town and Country Planning Act 1990 and The Town and Country Planning (Scotland) Act 1972. • Control of Noise (Appeals) Regulations 1975 (SI 1975 No. 2116). • Control of Noise (Measurement and Registers) Regulations 1976 (SI 1976 No. 37). • Noise at Work Regulations 1989 (SI 1989 No. 1790). • Road Vehicles (Construction and Use) Regulations 1986 (SI 1986 No. 1078) (various amendments). • Control of Noise (Code of Practice for Construction and Open Sites) Orders 1984 (SI 1984 No. 1992) and 1987 (SI 1987 No. 1730).
Policy and forthcoming legislation:	Some of the recommendations made by the Noise Review Working Party on legal control of noise were included in the Government White Paper *This Common Inheritance – Britain's Environmental Strategy*, concerning the reduction of the health risks arising from noise pollution by strengthening the law on noise. It was stated that this could be done by means of reducing noise at source, the implementation of standards, sound insulation requirements, and by planning and building regulation requirements. Consultation documents were issued in mid-1992 suggesting amendments to both the EPA and the COPA. A draft Planning Policy Guidance note is presently out to consultation (PPG 23). A private Bill on Noise and Statutory Nuisance has been proposed. The draft Bill includes provision that noise caused by a vehicle, machinery or equipment in the street is a statutory nuisance under the EPA. At the end of 1992 the EC proposed a Directive on the minimum health and safety requirements regarding the exposure of workers to the risks arising from physical agents. The proposed Directive specifically covers noise as one of four physical agents. The proposal contains minimum health and safety requirements and includes provisions for risk assessment and the provision of personal protection equipment. Annex 1 sets threshold and ceiling levels of noise.
Policy references:	• *Report of the Noise Review Working Party* (the BATHO Report), 1990. • Government White Paper: *This Common Inheritance – Britain's Environmental Strategy*, 1990. • Government White Paper: *This Common Inheritance – Britain's Environmental Strategy, Second Year Report*, 1992. • DoE Circular 10/73, *Planning and Noise*. • DoE Circular 2/76, Control of Pollution Act 1974 Implementation of Part III. • DoE Circular 8/81, The Local Government Planning and Land Act 1980, various provisions. • DoE Draft Planning Policy Guidance (PPG 23), Planning and Noise, 1992. • *Bothered by Noise? A guide to noise complaints procedure*, DoE, Welsh Office and Scottish Office, 1992. • DoE, *A guide to noise abatement zones*. • HSE, *Noise in Construction*, 1992. • HSE, *Noise at Work – Guidance on Regulations*, 1989. • *Exposure of Construction Workers to Noise*, CIRIA Technical Note 115, 1984. • *Simple Noise Screens for Site Use*, CIRIA Special Publication 38, 1985. • EC draft Directive on the Minimum Health and Safety Requirements regarding the Exposure of Workers to the Risks arising from Physical Agents (COM(92) 560 final), 1992. • Noise and Statutory Nuisance Bill (as amended by Standing Committee D, 1993).

Issue C2.1.6	Health and Safety at Work etc Act, 1974

Current legal position:	The Health and Safety at Work etc. Act 1974 (HSWA) introduced an entirely new system for the control of hazards in the work place and founded an independent Health and Safety Commission. The Health and Safety Commission has responsibility for developing legislation and policy, with a separate Health and Safety Executive which carries out operational and policing functions. Sections 2 and 3 of the HSWA impose on employers a general duty to ensure, so far as is reasonably practicable, the health, safety and welfare at work of all employees and of others who may be affected by the conduct of their undertakings. The HSWA itself does not contain specific legal obligations, only general duties. However, detailed regulations and codes of practice have been implemented under the Act. In practice these provisions require careful scrutiny of the measures necessary to deal with risks to health. This legislation also provides the basis for controls over hazardous substances, their development, storage and marking. The HSWA applies to all employers, employees and self-employed. It also applies to persons in control of non-domestic premises; manufacturers and supplier of commercial products and dangerous substances. Section 7 imposes a duty on employees to take reasonable care for their own safety and that of their employees. In relation to industrial activities, the HSWA is enforced by inspectors of the Health and Safety Executive. Local authorities' environmental health officers enforce legislation in relation to offices and shops. The penalties for most offences under the HSWA are a fine of a maximum of £20,000 for a summary conviction in the Magistrates' Court. In the Crown Court, on conviction or indictment, there is an unlimited fine and/or (for certain serious offences) imprisonment of up to two years. There are many regulations which have been made under the HSWA, the most important of which are listed below. Such regulations largely deal with occupational health and safety but in some cases they are important in an environmental context.
References to the current legal position:	**UK** • Health and Safety at Work etc Act 1974. • Management of Health and Safety at Work Regulations 1992 (SI 1992 No.2051). • Control of Asbestos at Work Regulations 1987 (SI 1987 No.2115) (as amended by SI 1992 No.3068). • Noise at Work Regulations 1989 (SI 1989 No.1790). • Ionising Radiation Regulations 1985 (SI 1985 No.1333). • Reporting of Injuries, Diseases and Dangerous Occurrences Regulations 1985 (SI 1985 No.2023) (as amended by SI 1989 No.1457). • Control of Industrial Major Accident Hazards Regulations 1984 (SI 1984 No.1902) (as amended by SI 1988 No.1462 and SI 1990 No.2325). • Control of Lead at Work Regulations 1980 (SI 1980 No.1248).
Policy and forthcoming legislation:	The proposed Construction (Design and Management) Regulations will impose duties on clients, designers, planning supervisors, principal contractors at all stages from project formulation and development through to the post construction, handover phase. The aim is to ensure the effective direction and coordination of health and safety throughout a construction project.
Policy references:	• In respect of the regulations listed under current legal position approved codes of practice or guidance is issued by the Health and Safety Commission. See Publications In Series, List of HSC/E publications (twice yearly). • HSE, *Construction Summary Sheets* (Ref SS) for small contractors covering many aspects of health and safety in the construction industry. • Health & Safety Commission, *Proposals for Construction (Design and Management) Regulations and Approved Code of Practice*, 1992.

Issue C2.1.7	Control of Substances Hazardous to Health Regulations

Current legal position: ▸ C2.1.6	The Health and Safety at Work etc. Act 1974 empowers the Secretary of State for Employment to make regulations concerning the health and safety of employees whilst at work. One such set of regulations is the Control of Substances Hazardous to Health Regulations 1988 (COSHH). The aim of these regulations is to protect against risks, either immediate or delayed which arise from exposure to substances which are hazardous to health. The list of substances which are defined as 'hazardous to health' includes: • substances which are toxic, harmful, corrosive or irritants; • substances for which the regulations specify a maximum exposure limit; • dust which is present in substantial airborne quantities; • micro-organisms which are hazardous to health; • and any other substance which creates a risk to health. The regulations do not apply to individual substances covered by other regulations e.g asbestos and lead; or substances which are hazardous to health merely because they are radioactive, explosive or flammable. The basic duty is to prevent or, where prevention is not practicable, adequately control the exposure of employees (and any other person who might be affected by the work) to hazardous substances. The key elements of the COSHH regulations are: • the assessment of the health risk associated with work involving hazardous substances; • the control of exposure to the substances hazardous to health; • the monitoring of exposure to substances hazardous to health; • the conducting of health surveillance; • the provision of information, instruction and training; • the prohibition of the use of certain substances.
References to the current legal position:	• Health and Safety at Work etc. Act 1974. • Control of Substances Hazardous to Health Regulations 1988 (SI 1988 No.1657) (as amended by SI 1990 No.2026, SI 1991 No.2431 and SI 1992 No.2382).
Policy and forthcoming legislation:	There are proposals for further amendments to the COSHH Regulations (HSC Consultation Document CD55 and CD56), 1993. Draft Approved Code of Practice – Control of Respiratory Sensitisers (HSC Consultation Document CD50), 1992.
Policy references:	• *COSHH General Approved Code of Practice* (4th Edition 1993). • HSE, *A Step by Step Guide to Assessment* (Ref HS(G) 97), 1993. • HSC, *The Control of Substances Hazardous Health in the Construction Industry*, CONIAC, 1990. • HSE, *Occupational Exposure Limits*, Guidance Note EH40/90, 1990. • HSE, *Hazard and Risk Explained*, COSHH Regulations (IND(G) 67(L), 1989. • HSE, *COSHH: A Brief Guide for Employers* (IND(G) 136(L), 1993. • *COSHH in Construction*, A Building Employers' Confederation Guide. • HSE, *Protection of Workers and the General Public During the Development of Contaminated Land* (Ref HS(G) 66). • *A guide to the control of substances hazardous to health in design and construction*, Report 125, CIRIA, 1993.

Issue C2.1.8	Contaminated land legislation and policy

Current legal position: ▸ C2.1.1 ▸ C2.1.2 ▸ C2.1.3 ▸ C2.1.6/C2.1.7 ▸ C1.1	The issue of contaminated land is important because of the risks to human health and to building integrity. Legislation in respect of contaminated land is derived from several statutes. Most of that legislation addresses either directly or indirectly, legal obligations and liabilities arising in respect of contaminated land e.g. under the Environmental Protection Act 1990 (EPA) (See C2.1.1 for general provisions of the EPA). However, there are no legislative standards set for clean up of contaminated land per se although there is government guidance. The following specific liabilities should be considered by a purchaser or developer of land: • *Section 79–82 EPA statutory nuisance* – liability for any accumulation or deposit which is prejudicial to health or a nuisance; • *Section 33 EPA waste management licensing provisions* – the treatment of contaminated soil on site, unless it is excepted by regulations, will require a waste management licence; • *Section 34 EPA* – duty of care regarding waste; • *Water Resources Act 1991* – liability may arise for reimbursement of costs incurred by the regulatory authority in respect of works they considered necessary to carry out in order to prevent poisonous, noxious or polluting matters from entering controlled waters; • *Occupiers' Liability Acts 1957 and 1984* – there is a duty to take such care as is reasonable in the circumstances to ensure the safety of persons (visitors and trespassers) entering a site. Liability will ensue for the owner or occupier if that duty is broken; • *Town and Country Planning legislation* including the Assessment of Environmental Effects Regulations 1988 (SI 1988 No. 1199 as amended) – contaminated soil may be a material consideration for the purpose of granting planning permission. If it is known about at the time of a grant of planning permission it is likely to be the subject of a planning condition or a planning agreement; • *Health and Safety legislation* – developers will have obligation to workers on site under the Health and Safety at Work etc Act 1974 and the COSHH Regulations; • *Building Regulations 1991* – regarding precaution to be taken in order to avoid danger to health caused by substances on or in the ground which is to be covered by a building. Liability for contaminated land can also accrue under common law. Anyone who suffers damage as a result of the presence or the escape of substances from contaminated land may commence a civil action under the common law rules of nuisance, trespass, negligence or the rule in Rylands –v– Fletcher. These are explained in more detail in C1.1.
References to the current legal position:	• Environmental Protection Act 1990 and related regulations (see C2.1.1). • Water Resources Act 1991. • Occupiers' Liability Acts 1957 and 1984. • Town and Country Planning Acts, The Town and Country Planning (Scotland) Act and related regulations. • Health and Safety at Work etc Act 1974 and related regulations. • Building Act 1984 and the Building Regulations 1991.

Issue C2.1.8	Contaminated land legislation and policy

Policy and forthcoming legislation:	The only government guidelines are from the Interdepartmental Committee on Redevelopment of Contaminated Land (ICRCL) which acts as an advisory body to the DoE and those wishing to redevelop contaminated land. The ICRCL issues guidance notes on aspects of contaminated land and their remediation. The ICRCL uses the concept of 'trigger concentrations', depending on the intended use of the site ('end use') to assist in determining the significance of contamination and to set acceptable concentrations of contaminants for the different end uses. Remedial methods are also outlined. The guidance is at both a general level and more specific e.g. relation to development of gasworks, scrap yards. The recommendations of the ICRCL may assume the status of a legal requirement if incorporated into a planning agreement. Dutch and American soil clean-up standards are sometimes referred to and may be useful as a comparison to the ICRCL guidance notes.
	Approved Documents – the guidance issued by the DoE in respect of the Building Regulations 1991 – list types of sites likely to contain contaminants. They also refer to lists of potential contaminants and state the recommended action to be taken in respect of them.
▸ D3.2.1	Under the planning system development, plans may set out policies for the reclamation and use of contaminated land.
▸ D3.2.2	There are several areas of legislation either to be enacted or yet to come into force, of relevance to contaminated land issues.
▸ D3.2.3	• *Section 61 EPA* – A date has not yet been set for the implementation of this section. If and when it comes into force, the section will allow a waste regulation authority to recoup any clean-up costs incurred by it where it has undertaken works necessary to avoid pollution occurring at a closed landfill site.
	• Although the proposed registers of contaminative uses have now been abandoned (Section 143 EPA), the DoE is to conduct a wide-ranging inter-departmental review of the problems of land pollution and the use of contaminated land.
	• *Environmental Assessment Regulations* – The classes of development to which the regulations relate may be extended.
	• *Liability for remedying environmental damage* – There is a proposal at EC level to impose strict liability for environmental damage howsoever caused. A Green Paper has recently been issued.
Policy references:	• DoE, Approved Document C to the 1991 Building Regulations, 1992. • DoE, *Construction of New Buildings on Gas Contaminated Land*, 1991. • DoE Circular 21/87, Development of Contaminated Land. • DoE Circular 17/89, Landfill Sites: Development Control. • DoE, Draft Planning Policy Guidance (PPG 25), Planning and Pollution Control, 1992. • ICRCL Guidance Notes 59/83; 17/78; 18/79; 23/79; 42/80; 61/84; 64/85; 70/90, 1970–1990. • Venables, R.K et al, *Environmental Handbook for Building and Civil Engineering Projects, Volume 1: Design and specification*, CIRIA Special Publication 97, 1994. • HSE, *Protection of Workers and the General Public During the Development of Contaminated Land*. • CIEF, *Contaminated land – Insurance and liabilities in sale and transfer*, Notes of a meeting held on 14/9/93, CIRIA, 1993. • CIEF, *Contaminated land – Technologies and implementation*, Notes of a meeting held on 14/9/93, CIRIA, 1993. • Steeds, J.E., Shepherd, E. and Barry, D.L., *A guide to safe working practices for contaminated sites*, Unpublished CIRIA Core Programme Funders Report, FR/CP/9, July 1993, (in preparation as an open publication). • *Remedial treatment of contaminated land*, in 12 volumes, forthcoming CIRIA publications due to be published early 1994. • *Guidance on the sale and transfer of contaminated land*, Draft for open consultation, CIRIA, October 1993. • Institute of Environmental Health Officers, *Development of Contaminated Land – Professional Guidance*, 1989. • BRE, *Concrete in Sulphate-bearing Ground and Groundwaters*, 1981. • BS5930: 1981 *Code of Practice for Site Investigations*, BSI, Milton Keynes. • BSI, Draft for Development DD175: *Code of Practice for the Identification of Potentially Contaminated Land and its Investigation*, 1988.

Issue C2.1.9	Air pollution legislation and policy

| **Current legal position:** | There are several legal controls over air pollution both statutory and at common law. Most of the EC legislation relates to industrial plants and incinerators and the control of specific dangerous pollutants.

Environmental Protection Act 1990 (EPA)

• **Air pollution control** – The control regime in relation to air pollution in England, Wales and Scotland was changed significantly by the EPA. Industrial processes which result in aerial emissions are divided into Part A processes and Part B processes.
Part A processes are subject to the integrated pollution control regime (IPC) and all emissions are controlled by HMIP. Part B processes (which are regarded as less seriously polluting) are subject to the new air pollution control regime (APC) and come under local authority control. Both regimes require authorisation of the process by the relevant authority. A variety of conditions will be imposed automatically, as required by Sections 6 and 7 of the EPA, particularly on the manner in which the process is to be conducted and on what may be discharged, including quantity and quality limits on discharge. The most significant of these conditions is the requirement that 'best available techniques not entailing excessive cost' (BATNEEC) shall be used for preventing the release of prescribed substances or, where that is not practicable, reducing to a minimum and to render harmless any substances which are released. Conditions may also include prescribed emission limits, breach of which is a criminal offence.

The new regime came into effect on 1 April 1991 for all new processes and is being phased in by 1996 for existing processes. It is a criminal offence to operate a prescribed process without the relevant authorisation.

• **Statutory nuisance** – Dust, steam, smell or other effluvia from industrial, trade or business premises which are prejudicial to health, are defined as statutory nuisances under Section 79 of the EPA. In addition smoke (including soot, ash, grit), fumes and gases from any premises may also be a nuisance. The person responsible (in practice the contractor doing the work in question) can be required by the local authority to put a stop to the nuisance through an abatement notice. An aggrieved individual may apply to the magistrates for an abatement order. Breach of a notice or order is a criminal offence.

Alkali etc Works Regulation Act 1906 and Health and Safety at Work etc Act 1974 – Where existing processes are not yet subject to the new IPC or APC controls, they continue to be controlled by the air pollution control regime applying before the EPA. The Health and Safety at Work Act 1974 imposes a duty on persons having control of premises to use best practicable means for preventing emission into the atmosphere of noxious or offensive substances and to render them harmless. The provisions of the Alkali etc Works Regulation Act 1906, involve prior registration (not authorisation) before a process can be operated. The operator is under a duty to use the best practicable means to prevent the emission into the atmosphere of noxious or offensive substances, and to render harmless and inoffensive substances which are nevertheless emitted.

Clean Air Act 1993 – This consolidates the Clean Air Acts 1956 and 1968 and the amendments to them. The Act is the main legislative means of control over, smoke, grit, dust and fumes, and is generally enforced by local authorities. The main provisions of the Act co-exist with the statutory nuisance provisions of the EPA. The Act prohibits the emission of dark smoke from a chimney of any building. It does not, therefore, apply to bonfire smoke. Emissions lasting for short periods (as specified by regulations) are permitted.

The Act applies to emissions of dark smoke from industrial or trade premises which originate from sources other than chimneys. Demolition has been found to be a trade process for the purpose of the 1956 and 1968 Acts, and a bonfire on a demolition site emitting dark smoke could be found to be within the ambit of the 1993 Act. There are, however, regulations which exempt emissions from the burning of timber and most other waste resulting from demolition of a building or of a site and also emissions from tar, pitch, asphalt or other matter used in connection with surfacing. It is a defence to a charge under the Act to prove that contravention was inadvertent and that all practicable steps had been taken to prevent or minimise emissions.

As under the 1956 and 1968 Acts, smoke control areas (SCA) can be designated under the Act. A local authority may declare a SCA, or the Secretary of State may direct the local authority to submit proposals for a SCA to him for his approval. It is an offence in an SCA to allow smoke emissions from a chimney. |

Issue C2.1.9	Air pollution legislation and policy
Current legal position continued:	*Vehicle Emissions* – New vehicles (diesel, petrol-engined cars and light vehicles) must meet emission standards for carbon monoxide, hydrocarbons, oxides of nitrogen and smoke (diesel engines only). Emissions including particulates from new diesel-engined trucks are also regulated. For vehicles in service there is an annual roadworthiness check which includes an instrumented (rather than just visual) smoke check. There are regulations to ensure that HGVs are properly maintained.
References to the current legal position:	**EC** • Directive on air quality limit values for sulphur dioxide and suspended particulates 80/779/EEC (as amended). • Directive on a limit value for lead in the air 82/884/EEC (as amended). • Directive on air quality standards for nitrogen dioxides 85/203/EEC (as amended). • Directive on the approximation of the laws of member states relating to measures to be taken against air pollution by gas from positive ignition engines of motor vehicles 70/220/EEC (as amended) and the EC Directive relating to commercial vehicles 88/77/EEC (as amended). **UK** • Environmental Protection Act 1990. • Alkali etc Works Regulation Act 1906. • Health and Safety at Work etc Act 1974. • Clean Air Act 1993. • Road Traffic Act 1972. • Prescribed Processes and Substances Regulations 1991 (SI 1991 No 472) (as amended). • Clean Air (Emission and Dark Smoke) Exemption Regulations 1969 (SI 1969 No. 1262). • Dark Smoke (Permitted Periods) Regulations 1958 (SI 1988 No. 498). • Control of Industrial Air Pollution (Registration of Works) Regulations 1989 (SI 1989/318). • Air Quality Standards Regulations 1989 (SI 1989 No 317). • Health and Safety (Emissions into the Atmosphere) Regulations 1983 (SI 1983 No. 943) (as amended SI 1989 No. 319). • Control of Asbestos in Air Regulations 1990 (SI 1990 No. 556). • Road Vehicles (Construction and Use) Regulations 1986 (SI 1986 No. 1078) (as amended). • Motor Vehicle (Tests) Regulations 1981 (SI 1981 No. (594) (as amended).
Policy and forthcoming legislation:	The Government White Paper *This Common Inheritance – Britain's Environmental Strategy* commits the government to ensure that air quality standards are met and also to take action against growing vehicle emissions by tighter emission standards. Various guidance notes have been issued by the Secretary of State relating to both general procedures and individual prescribed processes subject to IPC and APC. The guidance notes for individual processes contain release levels into air, water and land, including release levels and techniques for the minimisation of all releases, compliance monitoring requirements and various other matters. The EC proposes to phase in stricter standards for the sulphur content of diesel fuel between 1995–1997.
Policy references:	• Government White Paper: *This Common Inheritance – Britain's Environmental Strategy*, 1990 and Second Year Report, 1992. • DoE Circular 69/68 Clean Air Act 1968. • DoE Circular 11/81 Clean Air. • DoE Circular 3/91 The Environmental Protection Act 1990 Part I. New Local Authority and Port Health Authority Air Pollution Functions. • DoT, *Guide to Maintaining Road Worthiness*, 1991. • DoE and DoT, *How to Report Smoky Diesels*, 1993. • DoE issues process guidance notes for IPC and APC. See IPR series for IPC and PG series for APC. There are also some general guidance notes GG series.

Issue C2.1.10	The Wildlife and Countryside Act 1981

Current legal position:	The Wildlife and Countryside Act 1981 has two main parts:
	• **Species Protection** – Part I deals with the protection of birds, animals, and plants. In general, in relation to birds and animals the offences relate to the killing, injuring or taking of any wild bird or any scheduled wild animal. There are further offences in relation to birds, of being in possession of a wild bird or its eggs, destroying a nest whilst being built or taking or destroying eggs. In relation to animals it is an offence to be in possession of a scheduled animal live or dead or any part of one, or intentionally damaging, destroying or obstructing places of shelter or protection of animals. In respect of plants it is an offence for anyone other than the owner or occupier of land to intentionally uproot any wild plant. It is a further offence to intentionally pick uproot or destroy any of the specified rare wild plants. (The Act also prohibits the sale or advertisement of wild animals and listed plants.)
	• **Habitat Protection** – Part II allows for habitat protection and nature conservation by the designation of sites which are of special scientific interest (SSSIs) by reason of the existence of any flora, fauna or geological or physio-graphical features. Notice of designation of an SSSI must be given to the local planning authority for the area where land is situated, to the owner or occupiers of that land and to the Secretary of State for the Environment. Representations may be made in respect of such notification. The Notice will specify the reasons why the land is of special interest and any operations that may damage that flora, fauna or those features. SSSIs enjoy special protection. If an owner or occupier intends to carry out a damaging operation on that land he must seek consent from English Nature (and/or its equivalents in others countries of the UK) or planning permission. The Act also allows for the designation of Nature Reserves and Marine Nature Reserves. It is through the Act that relevant international and EC obligations are met, for example, the designation of Special Protection Areas for Birds as required by the EC Directive on the Conservation of Wild Birds and the designation of wetlands under the Ramsar Convention.
	In addition, the NRA has statutory duties in relation to the conservation and enhancement of the aquatic environment.

References to the current legal position:	**International** • Convention on Wetlands of International Importance especially as Waterfowl Habitat (Ramsar Convention). 1971. **EC** • Directive on the Conservation of Wild Birds 79/409/EEC (as amended). • Directive on the Conservation of Wild Fauna and Flora 92/43/EEC (Habitats Directive) (intended to be implemented by 5 June 1994). **UK** • Wildlife and Countryside Act 1981 (as amended). • Natural Heritage Act 1991.

Policy and forthcoming legislation:	Government policy aims to integrate environmental and economic activity in rural areas, to conserve the landscape, habitats and species diversity and to give extra protection to special areas.

Policy references:	• Government White Paper: *This Common Inheritance – Britain's Environmental Strategy*, 1990. • Government White Paper: *This Common Inheritance – Britain's Environmental Strategy*, Second Year Report, 1992. • DoE Action for the Countryside, 1992. • DoE Circular 27/87: Nature Conservation. • DoE Circular 1/92: Planning Controls over Sites of Special Scientific Interest. • DoE Draft Planning Policy Guidance (PPG 24): *Nature and Conservation*, 1992. • DoE Planning Policy Guidance (PPG 7): *The Countryside and the Rural Economy* (revised 1992).

C2.2   Overall purchasing policies

Issue C2.2.1	Agreeing green purchasing policies
Background:	Irrespective of the extent of the environmental requirements of the contract, it may be appropriate at the project planning and contract letting stage to establish specific purchasing policies to cover environmental requirements, whether covering the specification or your own corporate environmental policy. It is difficult for contract buyers, who have a central role to play in meeting the environmental requirements of a project, to effectively take on additional requirements in mid-contract. So early agreement on the policies to be adopted, the criteria for deeming products to be environmentally acceptable, and the sources of information will be an invaluable help towards meeting those environmental requirements consistently and efficiently. A particular issue of concern here is the procedures adopted for also ensuring that sub-contractors' purchasing policies and procedures are as rigorous as those of the main contract. The British Board of Agrément undertakes the environmental assessment of construction products.
Background references:	• BS7750: 1992 *Specification for Environmental Management Systems*, BSI, Milton Keynes. • *Corporate environmental policy statements*, CBI, London, June 1992. • CIC Environment Task Group, *Our land for our children: an environmental policy for the construction professions*, Construction Industry Council, August 1992. • CIEC Environment Task Force, *Construction and the Environment*, BEC, May 1992. • Construction Industry Environmental Forum, *Green Clients*, Notes of Workshop held on 30/11/92, CIRIA, 1992. • Construction Industry Environmental Forum, *Purchasing and specifying timber*, Notes of meeting held on 23/02/93, CIRIA, 1993. • *Waste Recycling and Environment Directory*, Thomas Telford, London 1993.
Good Practice:	*When setting up a project team and agreeing the final contract:* • ensure that any ambiguities in the environmental requirements of the tender documents are clarified; • ensure that clear agreement is reached on which of any additional environmental requirements included in your tender are to be accepted; • consult any other contractors involved to coordinate environmental actions and responsibilities; • consider requesting from sub-contractors environmental policy statements and commitments to their implementation; • establish the role of any appointed environmental manager on the application of the agreed purchasing policy in practice; • establish agreed purchasing policies, including identifying responsibilities for environmentally based buying decisions, criteria for acceptability of materials and products, agreed sources of information and the basis on which you might choose to exceed the environmental requirements of the contract whilst meeting the performance requirements and your own cost requirements; • ensure that purchasing to avoid or minimise waste, for example, selecting appropriate lengths of timber, panelling, bolts etc, is also explicitly covered.
Good practice references and further reading:	• As background references, plus: • Hall, K., and Warm, P., *Greener Building Products & Services Directory*, Association for Environment Conscious Building Directory, Second Edition, 1993 • *Timber: Types and sources*, Publication L296, Friends of the Earth, 1993. • Croners, *Environmental Management*, with quarterly amendment service, Croner Publications Ltd, First Edition, October 1991 • Miller, S., *Going Green*, JT Design Build, Bristol. • Ove Arup & Partners, *The Green Construction Handbook – A Manual for Clients and Construction Professionals*, JT Design Build, Bristol, 1993. • Contact: British Board of Agrément: Tel: 0923 670844.

Issue C2.2.2	Awareness and training
Background:	The development of environmental concerns in the construction industry is relatively recent and many people in the industry remain unaware and/or sceptical of the progress towards a more environmentally responsible attitude. There is a clear need to take active steps through training and other initiatives to increase directors', managers' and other employees' awareness of the issues, concerns and opportunities available.
Background references:	• BS7750: 1992 *Specification for Environmental Management Systems*, British Standards Institution, Milton Keynes. • Construction Industry Environmental Forum, *Environmental issues in construction – A review of issues and initiatives relevant to the building, construction and related industries*, CIRIA Special Publications 93 and 94, 1993. • CIC Environment Task Group, *Our land for our children: an environmental policy for the construction professions*, Construction Industry Council, August 1992. • CIEC Environment Task Force, *Construction and the Environment*, Building Employers' Confederation, May 1992. • Halliday, S.P., *Building Services and Environmental Issues – The Background*, BSRIA Interim Report, April 1992. • Leicester County Council, *Building for the Environment: An Environmental Good Practice Checklist for the Construction and Development Industries*, Leicester County Council jointly with Leicester City Council, November 1992. • The Engineering Council, *Engineers in the Environment – Some Key Issues*, December 1990.
Good Practice:	*When setting up a buying team for a new project:* • review the extent of the environmental requirements of the contract and the extent to which they exceed the normal requirements of your own corporate environmental policy; • review the extent of the proposed team's experience and/or training in environmental matters and develop a programme to rectify any deficiencies related to the project environmental policy; • identify, from this document and the background references, a list of publications on environmental concerns, providing environmental guidance or providing information on environmental materials and products that should be purchased for the site buying team to use in helping to meet the project environmental policy.
Good practice references and further reading:	• Use the background references plus: • Venables, R.K et al, *Environmental Handbook for Building and Civil Engineering Projects, Volume 1: Design and specification*, CIRIA Special Publication 97, 1994. • The Engineering Council, *Engineers and the Environment: Code of Professional Practice*, 1993. • Halliday, S.P., *Environmental Code of Practice for Buildings and Their Services*, BSRIA, 1994. • Barwise, J., and Battersby, S., *Environmental Training*, Croner Publication, 1993. • See C2.2.3 for list of references on environmental information.

Issue C2.2.3	Information requirements
Background:	There is a comparative dearth of information available to contractors and their buyers on environmental aspects of purchasing. However, there are some useful directories and organisations from which practical advice can be obtained.
Background references:	No specific references identified.
Good Practice:	*At the outset of a new contract*, seek out the following references and contact the indicated organisations to provide the site team with as much relevant environmental information as possible to support their purchasing decisions.
Good practice references and further reading:	• Building Research Establishment, *CFCs in Buildings*, BRE Digest 358, 1992. • Construction Industry Environmental Forum, *Environmental issues in construction – A review of issues and initiatives relevant to the building, construction and related industries*, CIRIA Special Publications 93 and 94, 1993 and in particular, Project Report 9, the Bibliography. • Venables, R.K et al, *Environmental Handbook for Building and Civil Engineering Projects, Volume 1: Design and specification*, CIRIA Special Publication 97, 1994. • CIC Environment Task Group, *Our land for our children: an environmental policy for the construction professions*, Construction Industry Council, August 1992. • CIEC Environment Task Force, *Construction and the Environment*, Building Employers Confederation, May 1992. • Ove Arup & Partners, *The Green Construction Handbook – A Manual for Clients and Construction Professionals*, JT Design Build, Bristol, 1993. • Construction Industry Environmental Forum, *Green Clients*, Notes of Workshop held on 30/11/92, CIRIA, 1992. • Construction Industry Environmental Forum, *Purchasing and specifying timber*, Notes of meeting held on 23/02/93, CIRIA, 1993. • Croners, *Environmental Management*, with quarterly amendment service, Croner Publications Ltd, First Edn, October 1991. • *Timber: Types and sources*, Publication L296, Friends of the Earth, 1993. • *The Good Wood Manual*: Specifying Alternatives to Non-renewable Tropical Hardwoods, Friends of the Earth, January 1990. • *Timber: Eco-labelling and Certification*, Publication L295, Friends of the Earth, 1993. • Hall, K and Warm, P., *Greener Building Products & Services Directory*, Association for Environment Conscious Building Directory, Second Edition, 1993. • Elkington, J. and Hailes, J., *The Green Consumer Guide*, Guild Publishing, London, 1989. • Contacts: – Building Research Establishment: Tel: 0923 894040 Fax: 0923 664010 – The Building Centre: Tel: 071-637 1022 – Association for Environment Conscious Building: Tel: 0453 890757 – Building Services Research and Information Association: Tel: 0344 426511 Fax: 0344 487575 – Friends of the Earth: Tel: 071-490 1555 Fax: 071-490 0881 – CIRIA: Tel: 071-222 8891 Fax: 071-222 1708 – TRADA: Tel: 0494 563091 Fax: 0494 565487.

Issue C2.2.4	Sub-contractor management
Background:	A key issue for main contractors and project managers who have accepted environmental duties and responsibilities under a main contract is to ensure that their sub-contractors also honour their related duties and responsibilities to the environmental requirements of the specification and the main contractor's corporate environmental policy.
	In the same way that clients may mirror the development of quality management systems based on BS5750, with some requiring any contractor wishing to tender to have established a certified quality management system, main contractors may wish to consider imposing similar conditions on their sub-contractors. However, since BS7750 has only recently being piloted, and a revised version is awaited, even the most environmentally conscious contractors will not yet have BS7750-derived environmental management systems in place.
	Separately, as the later sections of this document demonstrate, the legal and regulatory environmental requirements on contractors and others involved on site are increasing. Main contractors should be explaining to prospective sub-contractors the principal environmental facets of the project and the legal requirements they will be expected to meet. In addition, if a formal environmental policy for the project has been agreed, or formal statements of specific environmental requirements drawn up, they should be included in tender documents sent to sub-contractors and a formal requirement placed on tenderers to make specific provision for meeting these requirements.
Background references:	• Miller, S., *Going Green*, JT Design Build, Bristol. • Ove Arup & Partners, *The Green Construction Handbook – A Manual for Clients and Construction Professionals*, JT Design Build, Bristol, 1993. • Construction Industry Environmental Forum, *Environmental issues in construction – A review of issues and initiatives relevant to the building, construction and related industries*, CIRIA Special Publications 93 and 94, 1993. • CIC Environment Task Group, *Our land for our children: an environmental policy for the construction professions*, Construction Industry Council, August 1992. • CIEC Environment Task Force, *Construction and the Environment*, Building Employers' Confederation, May 1992.
Good practice references and further reading:	• No references have been identified as giving specific guidance on how to include environmental matters in tender documents for building and civil engineering projects. However, the following guide gives cogent guidance on inclusion of health and safety requirements which will in time be extendable to environmental matters. • European Construction Institute, *Total project management of construction safety, health and environment*, Thomas Telford, 1992. • Construction Industry Environmental Forum, *Considerate builders and contractors*, Notes of a meeting held on 9/3/93, CIRIA, 1993. • BS7750: 1992 *Specification for Environmental Management Systems*, British Standards Institution, Milton Keynes.

Issue C2.2.4	**Sub-contractor management**
Good Practice:	*In drawing up tender documents for sub-contracts:* • include any agreed environmental policy statements and/or specific environmental requirements, and explain explicitly what is required of the tenderer to meet them; • whilst requiring all sub-contractors to comply with all current legislation that is applicable to the work, and without removing such a duty from the sub-contractors, consider including a list of relevant environmental legislation which the you wish to draw especially to the tenderers' attention; • consider drawing attention in any general provisions of the tender documents to common environmental features of items in the specification, such as exclusion of CFCs or the use of temperate rather than tropical hardwoods, which would otherwise only be discovered by tenderers as they studied the details; • draw attention to particular environmental features or impacts of the project and its location; • if appropriate, include guidance on how sub-contractors should demonstrate the commitment sought, for example through participation in considerate contractor schemes. *In drawing up a select tender list:* • consider, as environmental management systems are developed, the extent to which the client wishes to restrict his or her choice of tenderer to those firms that have an environmental management system in place or can in some other way demonstrate their commitment to environmentally responsible operations; • consult the designer(s) and client(s) on their wishes on this issue; • add environmental performance to the normal selection criteria of quality, financial position, and completion on time and to budget • seek from potential tenderers a copy of their corporate environmental policy statement and/or other documentation that demonstrates their experience of and commitment to environmentally responsible operations; • consider drawing attention to considerate contractor schemes and adding participation in such schemes to your list of environmental criteria on which sub-contractors may be selected. *In comparing tenders on environmental grounds and making a final selection:* • if not already part of your corporate environmental policy, agree at a senior management level the influence environmental considerations will, if any, have on the final selection of sub-contractors so that the selection criteria are clear before the comparison is started; • review the documentation on environmental performance submitted by the tenderers and rank them in environmental priority order for comparison with the other selection criteria rankings; • in that review, consider the relative importance to you and your client of, on the one hand, past actual performance of a stated environmental policy which may be limited in scope, and on the other, policy statements which may display a stronger environmental commitment but which have not yet been put into practice. *When agreeing the contract with the sub-contractor:* • ensure that any environmental requirements included in the tender documents are transferred to the contract documents; • similarly, ensure that any environmental issues included in their tender by the successful sub-contractor are also transferred to the contract documents; • check whether any additional environmental requirements have come to light since tender invitations were issued and consider whether and how they are to be covered in the contract; • if not already agreed, ensure agreement on measures for environmental monitoring of the sub-contractor's work during the contract; • ensure the appointment by the sub-contractor of an individual responsible for environmental liaison with your environmental manager as envisaged by BS7750; • ensure that the duty of care for waste is clarified with respect to waste removed on the sub-contractor's behalf by the main contractor, i.e. where a broker's role is anticipated.

► C2.1.3

C2.3 Green management of a site

Issue C2.3.1	**Need for and implementation of an environmental management system/BS7750**

Background: ▸ D1.3.2	BS7750 provides a specification for environmental management systems and, if your company has prepared an environmental management system in accordance with the standard, it will provide for the production of a project environmental plan in much the same way as a quality management system to BS5750 provides for the preparation of a project quality plan. Such systems, though not yet able to be independently certificated like quality systems, are likely to have a major impact on staff organisation through the appointment of an Environmental Manager. The addition of environmental performance as a target, in parallel with the addition of quality to time and cost in quality management systems or total quality management systems, will also be important.
	If your organisation has no environmental management system in place, it may well have adopted a Corporate Environmental Policy Statement (CEPS), and project or site managers will have a duty to their employer to implement that CEPS on their project to the best of their ability.
▸ C1.1 – C1.4 ▸ C2.2.1 – C2.2.4	Separately from your organisation's environmental standpoint, the contract may well impose specific or generalised environmental requirements on the contractor and sub-contractors, and active steps will need to be taken to meet those requirements.
▸ C2.3.2	In any of these cases, project and site managers will need to acquaint themselves fully with the environmental requirements of either the contract or their company, and will need to enhance awareness of these issues amongst their staff and workforce.
Background references:	• BS7750: 1992 *Specification for Environmental Management Systems*, BSI. • Miller, S., *Going Green*, JT Design Build, Bristol. • *Corporate environmental policy statements*, CBI, London, June 1992. • CIC Environment Task Group, *Our land for our children: an environmental policy for the construction professions*, Construction Industry Council, August 1992. • CIEC Environment Task Force, *Construction and the Environment*, BEC, May 1992. • Construction Industry Environmental Forum, *Environmental management in the construction industry*, Notes of meeting held on 22/09/92, CIRIA, 1992.
Good Practice: ▸ C2.3.2	*When planning a new project before going on site:* • if your organisation has an environmental management system in place, prepare the project environmental plan, including a list of the project's likely environmental effects and plans for the keeping of the project's environmental management records, and use the provisions and guidance in this checklist; • if your organisation has no environmental management system in place but has adopted a Corporate Environmental Policy, consider how that policy is to be translated into practice on your project, and use the provisions and guidance in this checklist; • if your organisation has neither an environmental management system nor a corporate environmental policy, review the provisions and guidance presented in this checklist guide, and draw up a plan of environmental action for your project to the best of your ability, whilst recognising that it cannot be expected to approach best practice without expressed corporate commitment to underpin it; • in any of these cases, include plans to raise awareness and provide appropriate training to site managers, other site staff and the workforce – see C2.3.2 below; • identify a project environmental manager (part- or full-time depending on the project's scale) and ensure he or she has written terms of reference and delegated authority.
Good practice references and further reading:	• Use the background references, plus: • Croners, *Environmental Management*, with quarterly amendment service, Croner Publications Ltd, First Edn, October 1991. • European Construction Institute, *Total project management of construction safety, health and environment*, Thomas Telford, 1992. • Halliday, S.P., *Environmental Code of Practice for Buildings and Their Services*, BSRIA, 1994.

Issue C2.3.2	Awareness and training
Background:	See C2.2.2.
Good Practice:	• Review the guidance given in C2.2.2 and apply it to the full site-based team, including the workforce. • Consider setting up, under the project's environmental manager, a team of interested site managers to assist in monitoring and developing the project's environmental performance, in a similar way to the role of quality teams or quality circles.
Good practice references and further reading:	As C2.2.2.

Issue C2.3.3	Recording of environmental performance during the construction phase
Background:	This should be part of normal good site management, and also a requirement of any environmental management system including any prepared to BS7750. The draft Construction (Design and Management) Regulations being developed by the Health and Safety Executive would also, if brought into force, require much more rigorous record keeping on health and safety matters than is normal at present and this may also influence the environmental records that are kept. Implementation of a total quality management system would also include effective record keeping on environmental matters.
Background references:	• BS7750: 1992 *Specification for Environmental Management Systems*, BSI. • Health & Safety Commission, *Proposals for Construction (Design and Management) Regulations and Approved Code of Practice*, 1992.
Good Practice:	*At project planning stage:* • ensure any project environmental plan developed under stage C2.3.1 above includes provision for regular monitoring of measurable aspects of the site's environmental performance, not only of the main contractor but of all sub-contractors; • prepare a register of environmental effects and make plans for its regular review and update, including identifying the individual responsible for maintaining it; • set up a log for environmental incidents and complaints, and assign responsibilities for dealing with such events including preparation and/or updating of procedures; • consider using appropriate systems of electronic and/or computer-based collection and storage for environmental data, particularly on hazardous materials, waste disposal etc.
Good practice references and further reading:	• As background references, plus: • European Construction Institute, *Total project management of construction safety, health and environment*, Thomas Telford, 1992.

Issue C2.3.4	Site-specific purchasing policies
Background:	See C2.2.1 on Agreeing green purchasing policies.
Good Practice:	*In planning the purchasing from site, project or site managers should:* • review the agreed purchasing policies, including the identified responsibilities for environmentally based buying decisions, the criteria for acceptability of materials and products, the agreed sources of information, and the basis on which you might chooses to exceed the environmental requirements of the contract whilst meeting the performance requirements and your own cost requirements; • set up procedures for their day-to-day implementation on site; • ensure that purchasing to avoid waste, for example, selecting appropriate lengths of timber, panelling, bolts etc, or ensuring minimum waste in ordering of concrete is also explicitly targeted; • confirm the role of any appointed environmental manager on the application of the agreed purchasing policy in practice.
Good practice references and further reading:	See C2.2.1.

Issue C2.3.5	Transport policies including selection and maintenance of site plant and other vehicles
Background:	There are five main areas of concern. • One of the most 'visible' current environmental issues, and one with much public awareness, is the matter of CO_2 emissions and 'pollution from cars and trucks'. Many organisations are therefore, from a financial as much as an environmental standpoint, progressively changing remaining petrol-engined fleets to diesel as and when it is cost-effective to do so. • Quite apart from the requirements of the new MOT Test, there is increased pressure to maintain vehicles, and particularly trucks and site plant, in a way which reduces exhaust emissions to a minimum. • A key environmental point for some people is the global resource implications of scrapping vehicles and plant before it is environmentally, rather than financially, appropriate. Although financial factors are likely to govern decisions in this area, if financial considerations are equal, extending the useful life of site plant and other construction vehicles through thorough and careful maintenance will be one way to demonstrate environmental commitment. • The environmental impact of the transport of materials and products used in construction, particularly civil engineering projects, could be reduced if more use were made of rail or sea freight, even occasional transport by canal barge or river and delivery by pipeline or conveyor. Cost is obviously a prime factor, and long-distance transport of materials is rarely cost-effective compared to local supply. However, environmental pressure groups believe more use could be made of alternative to road vehicles to transport construction materials when they *have* to be moved long distances, and it is perhaps too easy to assume that road transport should be used. • If road transport is be used, the correct compromise between load size, number of vehicle movements and the timing and impact of those movements is not always given active consideration.
Background references:	• CIEC Environment Task Force, *Construction and the Environment*, Building Employers' Confederation, May 1992.
Legal references:	• Motor Vehicles (Tests) Regulations, SI 1694, 1981 and subsequent amendments. (The 'MOT Test' regulations.)

Issue C2.3.5	Transport policies including selection and maintenance of site plant and other vehicles
Good Practice:	*In setting transport policies for new projects:* • if new vehicles or plant have to be purchased, add environmental performance to your normal selection criteria; • set up procedures for regular monitoring of the fleet's environmental performance and the rectification of excess exhaust emissions; • consider whether more use could be made than hitherto of rail transport, or even canals to transport bulk materials, particularly to large civil engineering projects; • when reviewing the use of road transport, actively consider the load size/trip frequency balance and make appropriate decisions for your project and the plant available to you; • consider bussing-in staff and operatives or using gang vans instead of personal transport; • in addition, when considering personal transport, consider whether journeys are *actually necessary*, and whether a simple telephone call, use of a fax or modem, or even developments such as video conferencing would not only be more environmentally responsible but financially advantageous as well; • consider the development of schemes for staff and workforce to encourage the reduction of fuel consumption of site plant and road vehicle fleets, and the possibility of matching bonus schemes.
Good practice references and further reading:	• CIEC Environment Task Force, *Construction and the Environment*, Building Employers' Confederation, May 1992. • Advisory Committee on Business and the Environment, *A guide to environmental best practice for company transport*, DoE, November 1992.

Issue C2.3.6	Minimising energy use
Background: ▸ C2.3.5 ▸ C3.9	The Construction Industry Employers' Council Environment Task Force, in its Report *Construction in the Environment* highlighted the area of energy efficiency and conservation as one of 5 key action areas. Although the energy used in buildings is much greater than in the construction process, the overall sum spent by construction firms on energy used on site was estimated to be substantial, in excess of £1 billion per year. The total cost of energy directly consumed – ie excluding energy used in the extraction of materials – may be as low as 0.3 % but may rise to over 15% for some civil engineering projects involving extensive earthmoving. The main elements of energy use on site – transport and site plant, and the site offices – are dealt with separately in C2.3.5 and C3.9.
Background references:	See C2.3.5 and C3.9.
Good Practice:	• Follow the guidance in C2.3.5 and C3.9. • In addition, look for improvements in energy efficiency in the operation of small tools, developing plans and procedures for ensuring they are turned off when not needed, kept well maintained, and use the most appropriate power source. Above all, remember that energy efficiency in such matters almost always has a financial benefit as well as an environmental one. • Recognise the conflicts that may arise, for example between minimising energy use and aiding site security through flood-lighting. • Include the benefits of minimising energy use into the training and awareness programmes developed under C2.3.2.
Good practice references and further reading:	• CIEC Environment Task Force, *Construction and the Environment*, Building Employers' Confederation, May 1992. • See C2.3.5 and C3.9.

Issue C2.3.7	Minimising water use
Background:	On sites where water is metered, it pays to keep consumption to a minimum commensurate with other sensible site practices such as dust control, but even on non-metered sites, there is wider environmental merit in reducing water consumption. In addition, if disposal of waste water involves, for example, the construction of disposal facilities, then reduced water consumption may reduce the scale of such works that are needed, with a matching reduction in cost.
Background references:	No specific references identified.
Good Practice: ▸ C3.9 ▸ C2.3.2	*In developing plans to manage site water supplies:* • adopt the guidance on water use in site offices given in C3.9; • include the benefits of minimising water consumption into the training and awareness programmes developed under C2.3.2; • plan carefully the location of water supply points; • put in place procedures for checking water pipes are not left running and checked for leaks on a regular basis; • even if not required by the water supply company, consider installing meters as a control and a means of checking on water consumption; • where possible, gather rainwater and used water in appropriate settling tanks for use on tasks where potable water is not required.
Good practice references and further reading:	No specific references identified.

Issue C2.3.8	Waste management, storage and re-use
Background: ▸ C2.1.2 ▸ C2.1.3	The current legal position on waste is set out in some detail in C2.1.2 and C2.1.3. It will be clear from that section that site management must take its duties and responsibilities for waste management seriously in order to comply with the legislation. In addition, management of waste from the site offices is a significant concern and is dealt with separately in C3.9 on the site office environment.
Background references:	See C2.1.2 and C2.1.3.
Good Practice: ▸ C2.1.2 ▸ C2.1.3	*In actively developing plans to manage waste on site:* • review C2.1.2 and C2.1.3 carefully, and the appropriate references if needed, and identify the particular wastes and responsibilities likely to occur on your new project; • identify who will be responsible for waste management on the site when operational, and ensure he or she has the necessary training and authority to ensure compliance; • identify licensed waste disposal contractors in the vicinity of the site and take appropriate steps to check their licences are valid and their operations meet your own standards in this sensitive area; • ensure adequate temporary storage of topsoil; • discuss your proposed procedures with the waste regulation authority; • identify any likely or possible hazardous waste and procedures for dealing with it; • check the carrier's registration with the issuing waste disposal authority and record the result; • carry out random checks on whether waste actually reaches its planned destination and record the result; • retain all records of waste disposal and storage for at least two years.

Issue C2.3.8	Waste management, storage and re-use
Good practice references and further reading:	• Building Employers' Confederation: *Business Bulletin enclosure: Disposal of waste, rubble and demolition waste*, 1992 *Consignment note for the carriage and disposal of building waste (including warnings and advice)*, April 1992 *Technical Factfile Checklist on the Environmental Protection (Duty of Care) Regulations 1991.*
Legal references:	See C2.1.2 and C2.1.3.

Issue C2.3.9	Strategies for dealing with sensitive areas such as archaeology, nature conservation, SSSIs
Background:	Construction sites may include features, both natural and man-made, which it has been agreed should be retained as part of the final works or development. Such agreements may have been negotiated with the planning authorities, with bodies such as the NRA, English Nature, English Heritage, RSPB or other environmental and conservation agencies, or with the client, and may have legal or contractual force behind them. The NRA has statutory duties in relation to the conservation and enhancement of the aquatic environment.
	Alternatively an understanding may have been reached with the local community or interest groups, or existing features identified for retention as part of the overall design philosophy for the site. These could include trees (particularly those protected by Tree Preservation Orders), water bodies, plant or animal species or habitats, landscape or archaeological features. Measures to protect such features will be needed during the construction process to ensure that they are not damaged in any way and agreements and understandings are not broken.
	Whatever the feature or method by which its retention has been arranged, strategies for dealing with the construction phase alongside the sensitive area must be developed and implemented.
Background references:	• *Environmental Assessment: A guide to the identification, evaluation and mitigation of environmental issues in construction schemes*, CIRIA Special Publication 96, 1993. • Construction Industry Environmental Forum, *Nature conservation issues in building and construction*, Notes of a meeting held on 23/3/93, CIRIA, 1993.
Good Practice: ▸ C3.6.1 ▸ C3.6.2	These issues should have been considered at the design stage. ***At the construction planning stage of a contract:*** • identify any sensitive areas in your project and, from the specification and other contract documents, identify any contractual obligations for their protection; • using the guidance provided in C3.6.1 and C3.6.2 later in this Handbook, develop practical plans to deal with the identified features and, in particular, consider how the phasing and timing of the necessary measures and the construction processes will be arranged to minimise the possibility of damage and/or disturbance to the sensitive areas; • disseminate to managers, staff and workforce not only the developed plans but also the importance attached to adhering to them; • maintain close liaison with the bodies involved in designating the sensitive areas to ensure their continued support; • make plans for dealing with unexpected discoveries such as an archaeological find.
Good practice references and further reading:	• Venables, R.K et al, *Environmental Handbook for Building and Civil Engineering Projects, Volume 1: Design and specification*, CIRIA Special Publication 97, 1994, sections D2.3, D3.2.2 and D3.5.3. • *Guidelines for minimising impact on site ecology*, RSNC, 1992. • Therivel, R. et al, *Strategic environmental assessment*, Earthscan Publications, 1992.

Issue C2.3.10	Sub-contractor management
Background: ▸ C2.2.4 ▸ C2.3.1	As indicated in C2.2.4, a key issue for main contractors and project managers who have accepted environmental duties and responsibilities under a main contract is to ensure that their sub-contractors also honour their related environmental duties and responsibilities. The purchasing policies for sub-contractors are dealt with in C2.2.4, and any environmental management system developed along the lines outlined in C2.3.1 will provide for the active environmental management of sub-contractors. Where no environmental management system is in place, what is needed are plans for the active transfer of environmental requirements to sub-contractors and the monitoring of their environmental performance.
Background references:	• Miller, S., *Going Green*, JT Design Build, Bristol. • Construction Industry Environmental Forum, *Environmental issues in construction – A review of issues and initiatives relevant to the building, construction and related industries*, CIRIA Special Publications 93 and 94, 1993 – Chapter 6, Legislation and policy issues. • CIC Environment Task Group, *Our land for our children: an environmental policy for the construction professions*, Construction Industry Council, August 1992. • CIEC Environment Task Force, *Construction and the Environment*, Building Employers Confederation, May 1992. • BS7750: 1992 *Specification for Environmental Management Systems*, British Standards Institution, Milton Keynes.
Good Practice:	*When setting up arrangements for environmental management of sub-contractors:* • ensure the procedures outlined in C2.2.4 are followed; • put in place procedures for the monitoring of sub-contractors' environmental performance and assign responsibility for undertaking the monitoring and securing corrective action when needed; • include an item about environmental matters on the agenda for regular progress meetings.
Good practice references and further reading:	• See background references, plus: • European Construction Institute, *Total project management of construction safety, health and environment*, Thomas Telford, 1992.

C2.4 Pollution control strategies

Issue C2.4	Pollution control strategies
Background: ▸ C2.6 ▸ C3.5.1 ▸ C3.5.2 ▸ C3.5.3 ▸ C3.5.4 ▸ C3.5.5 ▸ C3.7	The legal background is presented in section C2.1.2 – 5 and C2.1.9. At the project planning stage, it is important to develop project-specific pollution control strategies to cover a wide range of potential problems: • waste water control – see C3.5.1; • appropriate storage of materials; • avoidance of groundwater pollution – see C3.5.1; • disposal of other wastes – see C3.5.2; • air pollution, including dust and fumes, and its control – C3.5.3; • noise and vibration control – see C3.5.4; • traffic control – see C3.7; and • light control, particularly avoiding light nuisance to neighbours – see C3.5.5. These pollution problems will all be minimised by careful planning before site work starts, and through appropriate consultation with the relevant bodies – see C2.6.
Background references:	• Construction Industry Environmental Forum, *Water Pollution from construction sites*, Notes of meeting held on 20/10/92, CIRIA, 1992. • Curwell, S.R., March, C.G. and Venables, R.K., (Eds), *Buildings and Health, The Rosehaugh Guide to the Design, Construction, Use and Management of Buildings*, RIBA Publications, 1990, particularly, Chapters B1 on Dust and fumes, B2 on Gases, Vapours and Mists, B3 on Asbestos, B6 on Plastics, Resins and Rubber, and A2 on Assessment of Risk.
Good Practice:	*At the project planning stage:* • review the specification and construction plan to identify the potential problems that could be encountered on the site(s); • develop (as part of the Project Environmental Plan if your organisation operates an environmental management system) specific strategies for storage of materials and for avoiding and/or dealing with and monitoring performance on each potential pollution incident and write them up in procedures and plans that site management, staff and workforce can implement if needed; • develop contingency plans for the more serious potential problems which might, for example, affect neighbouring water courses; • consult C3.5.1 to C3.5.6 if further guidance is needed at this stage.
Good practice references and further reading:	• See background references, plus: • National Rivers Authority, *Pollution Prevention Guidelines Working at Demolition and Construction Sites*, NRA, July 1992.
Legal references:	See C2.1.2, C2.1.3, C2.1.4, C2.1.5 and C2.1.9.

C2.5　Contaminated land

Issue C2.5.1	**Client assurances and identification of contaminated areas on site**
Background:	As pressure increases to not develop greenfield sites and more 'brown field' sites are therefore developed, it is increasingly likely that building and civil engineering works will be constructed on previously-contaminated land. Whilst the investigation and treatment of such land is outside the scope of this handbook, it must be highlighted that contractors and others involved in the construction phase must be made aware of any contamination treatment that has preceded their arrival on site. This will enable them to be on the look out for residual contamination and unforeseen contamination that may have not been discovered during any surveys or treatment.
	There is an onus on clients and their professional advisors to inform contractors of past treatment, and on contractors to ensure that such information is forthcoming.
▸ C2.1.8	See C2.1.8 for a summary of the legal position on contaminated land.
Background references:	• Leach, B.A., and Goodger, H.K., *Building on derelict land*, CIRIA Special Publication 78, 1991. • Curwell, S.R., March, C.G. and Venables, R.K., (Eds), *Buildings and Health, The Rosehaugh Guide to the Design, Construction, Use and Management of Buildings*, RIBA Publications, 1990, Chapter 9 on Contaminated Land by Viney and Rees.
Good Practice:	*At the planning stage of the construction phase:* • the client's professional advisors should formally inform the contractor(s) in writing of any site investigation results showing contamination of the site and records of the treatment undertaken, so that the contractor(s) are fully informed and the likelihood of unforeseen contact with residual contamination is kept to a minimum; • they should also consider, if they have not done so already, the extent to which continued monitoring of, for example, the progress of biological treatment, is necessary, and discuss with the contractor(s) how such monitoring would interface with the construction processes; • the contractor's project managers must assure themselves that they have all the information they require to ensure as far as reasonably practical the safety of their workforce and others on site from the risks of exposure to contamination.
Good practice references and further reading:	• Steeds, J.E., Shepherd, E. and Barry, D.L., *A guide to safe working practices for contaminated sites*, Unpublished CIRIA Core Programme Funders Report, FR/CP/9, July 1993, (in preparation as an open publication). • *Remedial treatment of contaminated land*, in 12 volumes, forthcoming CIRIA publications due to be published early 1994. • *Guidance on the sale and transfer of contaminated land*, Draft for open consultation, CIRIA, October 1993. • Construction Industry Environmental Forum, *Contaminated land*, Notes of a meeting held on 12/1/93, CIRIA, 1993.
Legal references:	See C2.1.8.

Issue C2.5.2	**Plans to deal with any residual contamination and/or unforeseen contamination if discovered**
Background:	However well a site investigation is done, and however well remedial treatment of contamination is undertaken, some residual contamination is likely and there remains the possibility of unforeseen contamination being discovered during construction. Contractors and resident engineers and architects need to have in place contingency procedures for how to deal with such unforeseen contamination if it is discovered.
Background references:	See C2.5.1.
Good Practice:	*At the planning stage of the construction phase:* • contractors' project managers should prepare, in consultation with the client and/or his professional advisors, a contingency plan for dealing with the otherwise unforeseen discovery of contamination or supposed contamination, for example, the unearthing of an unidentifiable drum which might contain toxic chemicals.
Good practice references and further reading:	• Leach, B.A., and Goodger, H.K., *Building on derelict land*, CIRIA Special Publication 78, 1991. • Curwell, S.R., March, C.G. and Venables, R.K., (Eds), *Buildings and Health, The Rosehaugh Guide to the Design, Construction, Use and Management of Buildings*, RIBA Publications, 1990, Chapter 9: 'Contaminated Land', by Viney and Rees. • Construction Industry Environmental Forum, *Contaminated land*, Notes of a meeting held on 12/1/93, CIRIA, 1993. • Steeds, J.E., Shepherd, E. and Barry, D.L., *A guide to safe working practices for contaminated sites*, Unpublished CIRIA Core Programme Funders Report, FR/CP/9, July 1993, (in preparation as an open publication). • *Remedial treatment of contaminated land*, in 12 volumes, forthcoming CIRIA publications due to be published early 1994. • *Guidance on the sale and transfer of contaminated land*, Draft for open consultation, CIRIA, October 1993.
Legal references:	• See C2.1.8 for contaminated land. • See C2.1.2 and C2.1.3 for waste.

C2.6   Relations with relevant bodies and groups

Issue C2.6.1	**Appropriate consultation with NRA, planning authorities, environmental and conservation agencies**
Background:	Many construction projects will have a major impact on natural features of the landscape or on special features of the built environment. Many agencies and authorities have responsibilities and/or influence on how a construction project relates to these features and the liaison that will have occurred at the design stage will have to be extended and even developed during the construction phase.
Background references:	No specific references identified.

Good Practice:
 	
▸ D2.3
▸ D3.3
	At the planning stage of the construction phase:
	
	• identify from the specification, other contract documents and consultation with the engineer and/or architect any special environmental requirements of bodies such as:
	  – the National Rivers Authority (or its equivalents in Scotland or Northern Ireland);
	  – the relevant planning authority;
	  – environmental and conservation agencies such English Nature, English Heritage, National Trust, Civic Trusts, and RSPB (or their equivalents elsewhere in the UK);
	• identify the extent of liaison already established at the design stage;
	• identify the existing contacts in each organisation;
	• make plans to establish working relations with each appropriate organisation;
	• identify and assign responsibility to appropriate site staff to undertake the necessary liaison during the construction phase.

Good practice references and further reading:	• Early consultation should be made with the Planning Liaison Officer in the local area NRA office. Each NRA office holds information concerning Town and Country Planning liaison procedures. • Venables, R.K et al, *Environmental Handbook for Building and Civil Engineering Projects, Volume 1: Design and specification*, CIRIA Special Publication 97, 1994, Sections D2.3 on site issues and D3.3 on site layouts and environmental impacts. • *Environmental Assessment: A guide to the identification, evaluation and mitigation of environmental issues in construction schemes*, CIRIA Special Publication 96, 1993. • NRA, *Guidance notes for local planning authorities on the methods of protecting the water environment through development plans*, Draft May 1993, NRA, Bristol.

Issue C2.6.2	Relations with site neighbours and the local public
Background:	In developing any building or civil engineering site, the immediate local community will have a legitimate interest and concern in the size, shape and form of what is to be constructed, the working practices to be employed, the duration of the construction phase and who the project when completed is intended to serve. They may be concerned about a number of issues, of which property values and quality of life will almost certainly be of paramount importance. At the planning permission stage, they may have adopted a 'Not In My Back Yard' (NIMBY) stance against the project without any truly justifiable argument to back it up. On the other hand they may have had legitimate wide-ranging concerns and may have proved to be vehement adversaries to the project. If they feel that their views have been listened to and concerns met, they can be useful allies but, if they are ignored, they may have a seriously adverse effect on the project's timescale and cost.
Background references:	• Construction Industry Environmental Forum, *Environmental issues in construction – A review of issues and initiatives relevant to the building, construction and related industries*, CIRIA Special Publications 93 and 94, 1993, Chapter 7 – Planning land use and conservation. • CIEC Environment Task Force, *Construction and the Environment*, Building Employers' Confederation, May 1992.

Good Practice:
 	
▸ C3.3.2
 	
▸ C2.6.1
	At the planning stage of the construction phase:
	
	• identify from consultation with the engineer and/or architect any special consultation undertaken at the design stage with local interest groups and identify any specific requirements of such groups which may have been incorporated into the design;
	• identify the extent of formal liaison already established and identify the existing contacts in each group or organisation;
	• consider joining a considerate contractor scheme if one is operating in the locality – see C3.3.2;
	• consult local resident's associations, pressure and action groups as appropriate, the Parish Council(s) affected and the Citizens' Advice Bureaux;
	• undertake such consultation in an atmosphere of trust rather than confrontation;
	• as with the agencies mentioned in C2.6.1, make plans to establish working relations with each appropriate group or organisation identified here;
	• consider the recommendations of the Environmental Law Foundation which include:
	  – writing to local people likely to be affected by the opening up of a construction site giving an estimated completion date, explanations of the uncertainties, a contact name, address and telephone number
	  – ensuring there is an informative notice on the site displaying the above information
	  – ensuring promised limits on hours of working are adhered to
	  – making sure those affected are informed about major deliveries or other major traffic movements
	  – ensuring that vehicles' wheels are cleaned before leaving the site;
	• make presentations to explain the effects of any mitigation measures planned and/or about the traffic movements to, from and around the site;
	• identify and assign responsibility to appropriate site staff to undertake the necessary liaison during the construction phase;
	• consider establishing means of post-construction appraisal of the site's impact on the local environment.

Good practice references and further reading:	• Construction Industry Environmental Forum, *Considerate builders and contractors*, Notes of a meeting held on 9/3/93, CIRIA, 1993. • *Considerate Contractor Schemes Explained*, Corporation of London, undated.

C2.7 Special considerations for design and build projects

Issue C2.7	Special considerations for design and build projects
Background: ▸ D1.1 ▸ D1.2 ▸ D1.3.2	On design and build projects the contractor will be taking on responsibility for the detailed design of the scheme, and in some cases for the primary design as well. On such projects the contractor needs to be aware of the environmental policy established by, or agreed, with the client and to communicate all relevant information to the members of his own design and construction teams. The environmental policy should be set out in tender documents, but there may be opportunities to explore this further with the client or suggest improvements to the selection of materials and/or products and methods of construction which reduce the environmental impact of the project.
Background references:	No specific references identified.
Good Practice: ▸ D1.1 ▸ D1.2 ▸ D1.3.2	*When agreeing responsibilities:* • a key member of the team should be appointed to ensure implementation of the environmental policy as envisaged in BS7750; • each member of the design and construction team should have a copy of the project environmental policy and be aware of its implications; • all members of the team should endeavour to keep abreast of changes in legislation, new information and environmental trends; • reference should be made to the design stage handbook where appropriate. *In design and site co-ordination:* • close liaison with the original design consultants is to be encouraged and where necessary they should attend meetings with the contractor's design and construction teams to explain key issues; • check whether additional environmental requirements have come to light since taking over the responsibility for project design and construction and refer back to the client for his consideration as to whether they should be incorporated within the contract; • ensure there is agreement with site staff on measures for environmental monitoring of work on site.
Good practice references and further reading:	• Miller, S., *Going Green*, JT Design Build, Bristol, undated. • Venables, R.K et al, *Environmental Handbook for Building and Civil Engineering Projects, Volume 1: Design and specification*, CIRIA Special Publication 97, 1994. • Ove Arup & Partners, *The Green Construction Handbook – A Manual for Clients and Construction Professionals*, JT Design Build, Bristol, 1993.

Stage C3   Site set-up and management

Stage 3: Site set-up and management covers site-specific planning, using the principles outlined in Stage 2, plus general guidance on environmental management of the site on a day-to-day basis.

C3.1   Legislation and policy

C3.2   Positioning, layout and planning of the site compound

C3.3   Relations with site neighbours and the local public

C3.4   Implementing green management plans on site

C3.5   Implementing pollution control strategies

C3.6   Protection of sensitive areas of the site

C3.7   Traffic management

C3.8   Environmental impact of temporary works

C3.9   Environment in the site offices

C3.10   Special considerations for specific civil engineering projects

C3.1   Legislation and policy

Issue C3.1	Legislation and policy
Current legal position: ► C1.1 ► C2.1	During the stage of site set-up and management several environmental legal requirements may apply. These legal requirements do not relate specifically to this stage in the construction phase but are of a more general nature and application. Sections C1.1 and C2.1 give an overview of the areas of relevant environmental legislation and policy. In most cases the relevant environmental consent and permits, for example relating to discharges and noise control, should have been granted before site set-up and management. In particular: • noise control may be an issue depending on the equipment used and C2.1.4 sets out the legal requirements. Under the Control of Pollution Act 1974, a developer or contractor may request that the local authority identify noise requirements in advance of the construction phase commencing; • discharges to controlled waters such as rivers, or groundwater or to the sewerage system are controlled under the water legislation: consents may be required – see C2.1.5; • any waste which is produced, carried, kept or disposed of will be subject to the duty of care as respects waste (see C2.1.3), the waste management provisions may also be applicable and a waste management licence may be needed – see C2.1.2; • the general requirements of the Health and Safety at Work etc Act 1974 and related regulations will apply to regulate the exposure of employees and subcontractors responsible for site set up and management to health and safety risks during these operations – see C2.1.6 and C2.1.7.
References to the current legal position:	• For references to the current legal position, please refer to the references quoted in the issue numbers set out in current legal position above.
Policy and forthcoming legislation:	No policy or forthcoming legislation issues for this stage have been identified although the Construction (Design and Management) Regulations, when finally issued in mid-1994, may affect the legal requirements on site managers.
Policy references:	• Health & Safety Commission, *Proposals for Construction (Design and Management) Regulations and Approved Code of Practice*, Health and Safety Executive, 1992.

C3.2   Positioning, layout and planning of the site compound

Issue C3.2	Positioning, layout and planning of the site compound
Background:	A civil engineering or building project may bring considerable benefits to the local community on completion or, conversely, may create significant impact. However, that is no reason to overlook the potential adverse impact, particularly on neighbours and local interest groups, of the site compounds, temporary though they are. Site managers need to be aware of this potential impact and position the compounds and activities within them such that nuisance to neighbours is minimised whilst efficient and effective working practices are still maintained.
Background references:	No specific references identified.
Good Practice: ► C3.5.5	Although it is possible that, on major projects, the location of the site compound will have been decided at design stage, on many projects its location is a matter for the contractor(s). To prevent conflict with the site's neighbours, care needs to be exercised in the positioning of the site compounds, particularly on large civil projects in rural areas where a badly-located compound and its many noisy activities can produce a much greater perceived nuisance response from the local community than the project itself. The local authority will have a major influence on access routes which may of course constrain the location beyond the contractor's control. ***In planning and designing the layout of the compounds, tackle four areas of concern.*** • **Noise survey prior to starting work** – It is advisable to undertake a noise survey prior to setting up the site compound(s) to provide a basis for comparison should there be any complaints about noise levels once construction is under way. • **Siting of activities in relation to neighbours and sensitive areas of the site** – Consideration should be given to siting the noisiest operations furthest away from immediate neighbours, to active screening of such operations by, for example, siting them behind site offices which might act as a noise screen, and to avoiding damage to sensitive areas of the site such as special habitats. Noise barriers (or visual barriers which also reduce noise) may be needed to reduce noise nuisance to neighbours. Equally, the layout of the lighting should, in addition to providing for efficient working and reasonable energy efficiency, be designed if at all possible to prevent intrusion of powerful lighting into neighbouring houses and other buildings. Site parking provision should, if at all possible, provide for all likely vehicles visiting the site to avoid extensive intrusion of site-related vehicles into the surrounding roads. If this is impossible, transport arrangements for staff and workforce should be considered. • **Drainage** – The siting of activities such as vehicle washing which generate large volumes of waste water must be carefully considered in relation to the drainage already available or needed to be installed, and consultation with the local sewerage provider and/or the National Rivers Authority may be essential. • **Deliveries and stores** – If at all possible, adequate space needs to be provided on site for delivery lorries and other vehicles to off-load into the stores area so that local roads are not at risk of being temporarily blocked. Careful planning of delivery times may also be necessary, and site staff given the authority to refuse deliveries attempted at the wrong time if the alternative is unacceptable congestion on neighbouring roads. The location of the stores compound needs careful planning not only to provide for efficient acceptance of deliveries and distribution to the site but also to minimise risks of intrusion by vandals and of exposure of neighbours to potentially hazardous materials and products.
Good practice references and further reading:	No specific references identified.

C3.3   Relations with site neighbours and the local public

Issue C3.3.1	Establishing good relations with neighbours and the local public
Background:	See C2.6.2
Good Practice: ▸ C2.6.2	If no plans were made at the contract letting stage for establishing good relations with neighbours and the local public, consult the guidance given under C2.6.2 and develop the specific plans mentioned there. Implementation of the plans may involve one, more than one or all of the following options: • public meetings to explain the construction phase of the project and its impact, for example the closure of roads, both permanent and temporary, particularly noisy stages such as piling, critical stages such as rolling in a bridge or installing major plant; • regular meetings with representatives of local groups; • an exhibition in a suitable local venue; • setting up liaison with local schools for visits and projects related to the contract; • on large, long-term projects, a Newsletter or regular bulletins on progress. In all this activity, be prepared for the cynical view that such measures are merely paying lip-service to environmental concerns and have your answer prepared.
Good practice references and further reading:	• *Considerate Contractor Schemes Explained*, Corporation of London, undated. • Construction Industry Environmental Forum, *Considerate builders and contractors*, Notes of a meeting held on 9/3/93, CIRIA, 1993.

Issue C3.3.2	Working hours and considerate contractors scheme
Background:	In addition to the problems of noise, lighting, dust and traffic, the timing of these operations (for example whether they take place at weekends) can also have a marked influence on the response of local community to a construction project. Considerate contractors schemes have also been set up by a number of local authorities based on provision of awards or commendations for tidy sites, establishing good relations with neighbours, considerate behaviour towards those affected by the site and care for the site's surroundings.
Background references:	• Construction Industry Environmental Forum, *Considerate builders and contractors*, Notes of a meeting held on 9/3/93, CIRIA, 1993. • *Considerate Contractor Schemes Explained*, Corporation of London, undated.
Good Practice: ▸ C3.3.1	Some projects will have as part of their planning permission restrictions on working hours and/or days and these not only need to be adhered to for legal reasons but the contractors' actions should also match the spirit of the restrictions in not placing too great a nuisance burden on the local community. *Where there are no imposed restrictions:* • be reasonable in relating the site's working hours to the proximity of neighbours to particularly annoying operations and to the specific nuisance they create; • if offering to restrict working hours on a noisy operation such as piling, avoid making promises you cannot live up to by, for example, offering never to drive piles after 6pm when the technical specification may require a pile to be driven to refusal and refusal has not been reached by 6pm – restarting the pile-driving the following day may be impossible; • where an operation requires a continuous period of working that is longer than the normal working day, use the public consultation liaison set up under C3.3.1 to inform and explain the operation to neighbours and why it is necessary; • note that the local authority and community *may* accept long working hours and weekend working if the overall duration of the project may be reduced, but avoid making unrealistic promises. Separately, consider joining a Considerate Contractor Scheme if one is operating in the locality of the site. Those identified involve membership on a site-by-site basis rather than enabling a contractor to register for all his work in a borough and require the contractor to abide by a code of practice and be inspected on a regular basis.
Good practice references and further reading:	• See background references, plus: • *Guidance Notes for Activities on the Public Highway in the City of London*, City Engineer's Department, Corporation of London, 1990.

C3.4 Implementing green management plans on site

Issue C3.4	Implementing green management plans on site
Background: ▸ C2.2.2 ▸ C2.3.1 ▸ C2.2.3 ▸ C2.3.5 ▸ C2.3.6 ▸ C2.3.7 ▸ C2.3.8 ▸ C2.3.9 ▸ C3.5	Section C2.3 provides guidance on preparing plans to tackle a number of issues involved in the active environmental management of a site which need to be addressed at the construction planning stage. They are: • the need for and/or implementation of an environmental management system (see C2.3.1); • awareness of and training on environmental issues (see C2.2.2); • monitoring and recording of environmental performance (see C2.3.3); • transport policies (see C2.3.5); • minimising energy use (see C2.3.6); • minimising water use (see C2.3.7); • minimising waste, and waste management (see C2.3.8); • strategies for dealing with sensitive areas such as SSSIs (see C2.3.9). At this stage, we must consider how such plans are to be implemented. Implementation of pollution control strategies is covered in C3.5.
Background references:	See separate issues in C2.3.
Good Practice: ▸ C2.3	***Implementation generally*** • If you are now on site and no plans along the above lines are available to you, refer to Section C2.3 and develop your own plans to suit your site and stage in the construction process. • If such plans are available to you, study them carefully and, in consultation with their originator, refine them if necessary to take account of the actual situation on site as it exists now, referring to the guidance in C2.3 as required. • Include monitoring and reporting on environmental performance on the agenda of all site progress meetings. ***Environmental management systems*** • If your organisation has an environmental management system in place, familiarise yourself with its overall provisions as well as the project environmental plan, including roles and responsibilities of others in the company for environmental matters. • If your organisation has no environmental management system in place but has adopted a Corporate Environmental Policy Statement, familiarise yourself with the policy statement, try to discover how it has been implemented on other sites, and learn from their experience in what is a very new development in construction site management. • If your organisation has neither, do not divert yourself at site level to try to rectify that lack, but use the guidance in this handbook to actively manage the environmental performance of your site. • In any of these cases, identify the person who is to take on the role of project environmental manager (part- or full-time depending on the scale of the project) and ensure he or she has written terms of reference and delegated authority. ***Training and awareness*** • In implementing agreed training and awareness plans, recognise that active environmental management of construction sites is still a relatively new subject and awareness levels among many staff and the workforce may be low. • Recognise also that there are many environmental improvements, for example energy efficiency, that have a financial motivation as well as an environmental one and these and the legal requirements will provide useful entry points to raising awareness and changing behaviour. ***Information*** • Since practical environmental information is scarce, use this Handbook and the plans drawn up at contract letting stage to secure an appropriate library of environmental information on site to assist the environmental manager and others to be as well informed as practicable on the issues they face.

Issue C3.4	Implementing green management plans on site
Good Practice continued: ▸ C3.9	*Monitoring and recording of environmental performance* • Consider assigning this activity as high a priority as safety reporting. *Site-specific purchasing* • Consider making this the subject of a regular report at site management meetings as a way of maintaining commitment in this area. *Transport policies* • In seeking to reduce fuel consumption and improve efficiency, recognise that the *need* for a journey is at least as important a matter for serious consideration as other steps. • Recognise that well-maintained vehicles and other plant not only use less fuel than if maintenance is neglected but also cause less nuisance to other workers and neighbours, and last for a longer time. • Encourage all employees to adopt more considerate and less frantic driving styles, needless idling and revving of the engine. *Minimising energy use* • Consider making energy consumption the subject of a regular report at site management meetings as a way of maintaining commitment in this area. • This is a prime example of financial and environmental motives coinciding and, in view of the savings available – can be a most useful initial subject for introducing environmental matters to site personnel. *Minimising water use* • With metered supplies, the financial and environmental motives are much the same as for energy consumption. In addition, less effluent will be created for disposal. *Minimising waste, and waste management* • Although for some people energy and water may lead the list of greatest environmental concerns on site, waste is the certainly the most visible environmental issue of increasing concern and one on which obvious progress can be made. It is a cliche that a tidy site is an efficient site, but the experience of applying quality management systems to sites, especially to the control of materials and distribution from stores, bears out the linkage between the two. • At one level, recognise the new legal position and ensure, again through the regular site management meetings, that all necessary steps are being taken to comply. • On the other level, of conserving resources and minimising needless waste, recognise the financial motive here too and take steps to re-use to the full formwork, trench supports, and other re-usable temporary works, and try to engender a re-using culture amongst the site office staff and the workforce. Encourage feedback to those responsible for purchasing, for example, if plywood sheet sizes could be altered to reduce waste. *Sensitive areas* • Here, the need is to ensure the understanding and maintain commitment of staff and workforce to the long-term protection of certain areas of a site, and the acceptance that such protective measures may at times be significantly inconvenient and reduce site efficiency. *Sub-contractors* • Here, the main task is to ensure that the pressure to provide a good environmental performance is at least as firmly maintained on sub-contractors as it is on the main contractor.
Good practice references and further reading:	• Croners, *Environmental Management*, with quarterly amendment service, Croner Publications Ltd, First Edn, October 1991. • BS7750: 1992 *Specification for Environmental Management Systems*, British Standards Institution, Milton Keynes. • European Construction Institute, *Total project management of construction safety, health and environment*, Thomas Telford, 1992. • Skoyles, E.R. and Skoyles, J.R., *Waste Prevention on Site*, Mitchell Publishing, 1987.

C3.5   Implementing pollution control strategies

Issue C3.5.1	Waste water control

Background: ▸ C2.1.4	Construction sites always produce some waste water, sometimes in substantial quantities, as well as more obvious pollutants. For example, construction activities can affect the water environment in many ways, produce many sources of waste water or involve the containment or protection of ground water as follows: • the outflow from de-watering systems which may include silt, sediment and contaminated groundwater; • the run-off from washing ready-mixed concrete trucks and/or earth-moving equipment; • the hosing of dirt and concrete from various surfaces; • the leakage from oil and fuel tanks; • the spillage of oil or fuel through poor protection, vehicle damage or accidental valve opening – it is all-too-easy to hose away a spillage, say, of spilt oil or petrol without considering the consequences; • vandalism; • the dumping of construction debris into or near watercourses or surface water drains; • the accumulation of dirt and oil on hardstandings that is subsequently washed off during heavy rain; • the puncturing a natural impermeable layer by piling, thus permitting vertical movement of pollutants and leachates into underlying aquifers; • leachate pumping. In addition, particularly in remote civil engineering sites, there may be a need to set up run-off and/or sewage disposal facilities rather than simply connect to the local main drainage system. The disposal of these wastes needs to be carefully planned for and controlled. At risk are the local rivers and other fresh water, the ground water and, in more urban areas, workers at work in drains and other sewerage facilities who can so easily be overcome by the fumes from hosed away chemicals. The NRA is sympathetic to the problems faced by the construction industry and is available to assist construction organisations with the provision of information, as well as providing the guidelines listed below. However, it can call on legislation to prosecute anyone known to be causing a threat to the environment, so the threat of and actual prosecution can be expected to increase as part of the NRA's pollution control policy.
Background references:	• Construction Industry Environmental Forum, *Water Pollution from construction sites*, Notes of meeting held on 20/10/92, CIRIA, 1992.
Good Practice:	*In implementing waste water and water-based pollution control on site:* • use available guidance – see list below – consulting the local NRA office if needs be; • carefully site oil and fuel storage tanks, and provide appropriate protection measures such as bunds, oil and petrol separators or other secondary containment; • ensure secure valves are provided on oil and fuel supplies; • consider providing settling tanks or other separators for silt-laden material; • consider the level of site security appropriate to the pollution and other dangers; • seal off or remove abandoned drains to minimise the spread of contaminated water; • actively manage site surface water, for example by providing collection channels leading to oil and/or silt traps as appropriate; • consider using appropriate waste water for some site purposes to reduce consumption of mains water.

Good practice references and further reading:	• *River Pollution and how to avoid it*, National Rivers Authority Leaflet.
	• National Rivers Authority, *Pollution Prevention Guidelines:*
	  – *Disposal of sewage where no mains drainage is available*, March 1992
	  – *Above-ground oil storage tanks*, March 1992
	  – *Working at Demolition and Construction Sites*, July 1992
	  – *General Guide to the prevention of pollution of controlled waters*, July 1992
	  – *Works in, near or liable to affect watercourses*, July 1992.
	  – *The use and design of oil separators in surface water drainage systems*, 1992 •
	  Design of flood storage reservoirs, CIRIA/Butterworth, 1993.

Issue C3.5.2	Disposal of other wastes
Background: ▸ C2.1.2 ▸ C2.1.3	Section C2.1.2 deals in detail with the legislation and policy on waste, and C2.1.3 with the duty of care for waste. The prime concerns of site management on disposal of waste are: • to do what is necessary to comply with the law; • to do what is practicable and commercially sensible to minimise waste, to collect waste for recycling, and to find outlets for the sorted waste so that it may be re-used and/or turned into other useful products or materials.
Background references:	• *Environmental issues in construction – A review of issues and initiatives relevant to the building, construction and related industries*, CIRIA Special Publications 93 and 94, 1993, Chapter 6 on resources, waste and recycling. • CIC Environment Task Group, *Our land for our children: an environmental policy for the construction professions*, Construction Industry Council, August 1992. • CIEC Environment Task Force, *Construction and the Environment*, Building Employers Confederation, May 1992.
Good Practice: ▸ C2.1.2 ▸ C2.1.3	*To deal effectively with disposal of wastes:* • through existing company procedures (or if none, specially created procedures) ensure compliance with the legal provisions set out in C2.1.2 and 2.1.3; • ensure that any waste disposal contractors you use are appropriately registered for the waste you expect them to carry – check with the waste regulation authority and record your actions; • introduce 'good housekeeping' measures to control and minimise waste of materials on site, for example by keeping a tidy site, proper storage of materials, educating operatives to minimise waste, setting waste reduction targets; • have clear procedures to identify and record waste arising from changes or errors in design and construction; • be alert to problems arising from waste disposal including residual paints and solvents in containers, dusts from cement, timber and asbestos, and broken glass, all of which can cause safety hazards if not pollution problems; • consider whether it is feasible to segregate wastes and, if possible, classify site waste and separate it for salvage, re-use or recycling at the point of use; • take measures to avoid accidental contamination and mixing; • identify any 'special waste' and review disposal of surplus substances, and their containers, covered by the COSHH regulations; • if not already identified, search locally for disposal outlets for recyclable materials; • collect materials pallets for return to materials/products manufacturers; • administer necessary documentation with respect to duty of care.
Good practice references and further reading:	• DoE Circular 11/91: *Controlled Waste (Registration of Carriers and Seizure of Vehicles) Regulations*, 1991. • DoE Circular 4/92: *The Waste Recycling Payment Regulations*, 1992. • DoE Circular 14/92, *The Environmental Protection Act 1990 – Part II and IV: the Controlled Waste Regulations*, 1992. • DoE, *Waste Management: The Duty of Care – A Code of Practice*, 1991. • DoE Waste Management Paper No.1, *A Review of Options*, 1992. • DoE Waste Management Paper No.23 *Special Wastes: A Technical Memorandum Providing Guidance on their Definitions*, 1983. • DoE Waste Management Paper No.28 *Recycling: A Memoranda Providing Guidance to Local Authorities on Recycling*, 1991. • *A Guide to the Control of Substances Hazardous to Health in Construction*, Report 125, CIRIA, 1993. • There are several DoE Waste Management Papers dealing with specific wastes such as *Solvent waste* (No.14); *Wood preserving waste* (No.16); *Asbestos waste* (No.18).
Legal references:	See C2.1.2 and C2.1.3.

Issue C3.5.3	Air pollution and its control, including dust and fumes

Background:	It is acknowledged that some construction processes release pollutants to the atmosphere, for example dust from earthmoving. However, there is a clear need to keep them to a minimum to avoid undue nuisance to neighbours and the workforce, and to ensure that the more dangerous pollutants, for example paint sprays and sprayed formwork oils, are not released or are, at least, controlled.
	The main potential pollution generators to be considered here include:
	• bonfires; • brick and silica dusts; • toxic substances in the ground or services; • cement, fillers and plaster; • solvents, glues and paints; • weedkillers and other similar chemicals • some magnesium limestone dusts which are toxic to cattle if blown from sites onto adjacent fields.
▸ C2.1.9	The legal position is summarised in C2.1.9.
Background references:	• Government White Paper: *This Common Inheritance – Britain's Environmental Strategy* (1990). • Government White Paper: *This Common Inheritance – Second Year Report* (1992)

Good Practice:	*In implementing a site air pollution control strategy:*
	• relate the COSHH assessments undertaken on occupational health and safety grounds to the risks to the environment generally and neighbours in particular;
	• identify those potential pollutants and/or irritants that will need particular attention and control;
	• make and implement plans to control airborne pollutants such as:
	  – dust from earthmoving and other vehicle movements within the site
	  – wood dust from joinery and other woodworking
	  – fumes from welding
	  – sprayed formwork oils
	  – spayed paint and/or other sealants;
	• for example, consider applying water and polymer sprays to haul roads and using dust extractors on cutters and saws;
	• be aware of any bye-law restrictions on site fires and anyway limit, and if possible eliminate, the use of fires on site as a form of waste disposal;
	• secure all loose material that could be blown about and outside the site by a high wind, and make sure allowance is made for damp materials that will dry out in such circumstances.

Good practice references and further reading:	• Curwell, S.R., March, C.G. and Venables, R.K., (Eds), *Buildings and Health, The Rosehaugh Guide to the Design, Construction, Use and Management of Buildings*, RIBA Publications, 1990, in particular Chapters B1 Dusts and fumes, B2 Gases, vapours and mists, and B8 Fire. • *Manual on good sealant application*, CIRIA Special Publication 80, 1991 • Bielby, S. C., *Site Safety – A handbook for young construction professionals*, CIRIA Special Publication 90, 1992.
Legal references:	See references in C2.1.1 and C2.1.9.

Issue C3.5.4	Noise and vibration control
Background: ▸ C2.1.5	The legal position on noise pollution and its control is set out in some detail in C2.1.5 and, in environmental terms (as opposed to occupational health terms) the key concerns are: • to do what is necessary to comply with the law or to prevent enforcement by a regulatory authority or third party; • to do what is practicable and commercially sensible to minimise noise in order to maintain good relations with immediate neighbours and the local community. The potential structural effects of vibration on, for example, neighbouring properties should have been dealt with at the design stage and are considered by many not to be an *environmental* issue but a *structural* one.
Background references:	• Report of the Noise Review Working Party (the BATHO Report), 1990. • Government White Paper: *This Common Inheritance – Britain's Environmental Strategy*, 1990. • Government White Paper: *This Common Inheritance, Second Year Report*, 1992. • DoE Circular 10/73: *Planning and Noise*. • DoE Circular 2/76: *Control of Pollution Act 1974 Implementation of Part III*. • DoE Draft Planning Policy Guidance: *Planning and Noise*, PPG 23, 1992. • HSE, *Noise in Construction*, 1992. • *A guide to reducing the exposure of construction-workers to noise*, CIRIA Report 120, 1990.
Good Practice:	*To minimise noise nuisance to site neighbours:* • ensure all noise-related occupational health provisions are met, since this will as a consequence reduce the noise transmitted outside the site; • observe any operational restrictions imposed by the local Environmental Health Officer; • even if no operational restrictions have been imposed by the local Environmental Health Officer, keep working hours for particularly noisy operations to a minimum or, perhaps, with appropriate local consultation beforehand, extend the working hours of such operations to compress the nuisance to as few days as possible; • in blasting operations, sequencing the blasts can significantly reduce the resulting vibrations; • give due and proper notice of any blasting operations; • use the guidance on operational procedures listed below; • keep plant well maintained, and include noise performance checks in regular maintenance schedules.
Good practice references and further reading:	• HSE, *Noise at Work – Guidance on Regulations*, 1989. • *Exposure of Construction Workers to Noise*, CIRIA Technical Note 115, 1984. • *Planning to reduce noise exposure in construction*, CIRIA Technical Note 138, 1990. • *Simple Noise Screens for Site Use*, CIRIA Special Publication 38, 1985. • *Ground borne vibrations arising from piling*, CIRIA Technical Note 142, 1992. • BS5228: *Approved Code of Practice – Noise Control*, BSI, London. • BS4142 1990: *Methods for Rating Industrial Noise Affecting Mixed Residential and Industrial Areas* BSI, London. • Beaman, A. L. and Jones, R. D., *Noise from construction and demolition sites – measured levels and their prediction*, Report 64, CIRIA, 1977.
Legal references:	See C2.1.5.

Issue C3.5.5	Light control
Background: ▸ C3.2	As indicated in C3.2 on layout of the site compound, the layout of the compound and other site lighting should, in addition to providing for efficient working and reasonable energy efficiency, be designed if at all possible to prevent intrusion of powerful lighting into neighbouring houses and other buildings. If this is not done, the powerful lighting needed for sites to operate safely outside daylight hours can cause considerable nuisance, even preventing sleep if the lighting remains on all night for security reasons.
Background references:	See good practice references.

Good Practice:	*When designing and installing lighting of site compounds:*
	
	• consult the available guidance;
	• include consideration of the following key points:
	  – direct light downwards wherever possible;
	  – if that is not possible, try to use lighting designed to minimise light spread above the horizontal;
	  – if up-lighting is unavoidable use baffles to keep light spill to a minimum;
	  – take particular care in the positioning of floodlights to avoid light spill outside the compound straying into neighbouring property.

Good practice references and further reading:	• *CIBSE Lighting Guides* covering *The Industrial Environment, Hospitals and Health Care Buildings, Areas for Visual Display Terminals, Sports, Lecture Teaching and Conference Rooms*, and *The Outdoor Environment*, Chartered Institution of Building Services Engineers, in sections, 1989 – 1991. • *Guidance Notes for the Reduction of Light Pollution*, The Institution of Lighting Engineers, Rugby, 1992. • *A guide to the use of compact fluorescent lamps*, The Institution of Lighting Engineers, TR20.

C3.6   Protection of sensitive areas of the site

Issue C3.6.1	Trees, water, species, habitat and landscape features
Background:	Construction sites may include features which it has been agreed should be retained as part of the final development. Such agreements may have been negotiated with the planning authorities or bodies such as the NRA or English Nature and may have legal force behind them. Alternatively an understanding may have been reached with the local community or interest groups, or existing features identified for retention as part of the overall design philosophy for the site. These could include trees (particularly those protected by Tree Preservation Orders), water bodies, plant or animal species or habitats, landscape or archaeological features. Measures to protect such features will be needed during the construction process to ensure that they are not damaged in any way and agreements and understandings are not broken. The NRA has statutory duties in relation to the conservation and enhancement of the aquatic environment.
Background references:	See good practice references.
Good Practice: ▸ C2.1.4	*To ensure these features are practically and sensitively dealt with:* • consult English Nature and/or other relevant environmental and conservation organisations such as the London Ecology Unit for advice on how to deal with these sensitive areas in relation to the environment local to the site; • ensure all construction staff are made aware of the importance of protecting designated areas and why; • clearly identify, mark and/or fence off sensitive areas of a site – where necessary signs should be erected; • to avoid direct damage from vehicles or trampling, areas should be securely fenced off, preferably with chestnut paling or similar at least 4 feet high – less robust forms of marking or fencing off are generally inadequate; • where possible such areas should include all the features requiring protection plus a buffer zone – with trees this will mean any land covered by the root system, that is generally any area covered by the tree canopy; • protect water features and species or habitats from run-off from the construction site since they may be susceptible to damage from pollutants contained in such run-off – suspended solids can be just as damaging as chemical spills in certain situations; • take measures to ensure that run-off is away from sensitive areas and, where necessary, construct trenches or bunds to intercept any run-off (any waste water likely to reach a water course will need a discharge consent from the National Rivers Authority); • make one member of the team responsible for ensuring regular inspection and maintenance of and fencing, signs etc; • ensure any action is not taken at an inappropriate time, for example by not disturbing nesting birds or over-wintering wildfowl; • consider providing running boards over sensitive areas to prevent deep rutting; • take care in storing turf and topsoil; • if at all possible, choose routes for any haul roads that need to pass through such areas in a way that avoids damaging overhanging trees; • on major projects such as pipelines, take active steps to keep disturbance to a minimum, for example by setting up one-way systems for construction traffic to minimise turning and slewing, especially by tracked vehicles.
Good practice references and further reading:	• Construction Industry Environmental Forum, *Nature conservation issues in building and construction*, Notes of a meeting held on 23/3/93, CIRIA, 1993. • *Trees on Development Sites*, Arboricultural Association, 1985. • *Protected Trees: A guide to preservation procedures*, DoE and Welsh Office. • *Guidelines for minimising impact on site ecology*, The Royal Society for Nature Conservation, 1992.
Legal references:	See C2.1.1, C2.1.4 and C2.1.10.

Issue C3.6.2	Habitat translocation and/or creation
Background:	In some cases development may threaten the continued existence of some uncommon plant or habitat. This may arise directly as a result of construction activity, or indirectly from the intended use of the development. It may have been agreed with relevant parties that the best course of action is to move the affected species or area either off site altogether or to a better-protected part of the development site where hopefully it will continue to prosper. If this is so then it needs to take place before any major construction activity occurs which could potentially damage the species or habitats to be moved.
	Translocation is usually only agreed to by nature conservation organisations as an if-all-else-fails measure.
	Similarly, if the creation of new habitats has been agreed, you will need to consider how they will be achieved, although it may not be necessary or even desirable to carry out any particular work at this particular phase of the construction process.
Background references:	• Buckley, G.P. (Ed), *Biological Habitat Reconstruction*, Belhaven Press, London, 1989 (particularly Introduction and Philosophies of Habitat Reconstruction. pp.1–26.).
Good Practice:	• Review the guidance in C3.6.1.
	• The key to successful translocation is to have a thorough knowledge of the composition and requirements of the species or habitats being moved. Expert advice will be needed at all stages of the translocation procedure. A suitable receptor site needs to be identified which should comprise as many of the ecological characteristics of the donor site as possible to ensure a successful translocation. The receptor site will need careful preparation. Translocation can involve the movement of:
	– individual plants (including trees)
	– intact turfs;
	– layers of soil
	– seeds of plants.
	• In general the techniques employed are straightforward – usually the species or habitats are removed by hand or by using some form of mechanical shovel (possibly adapted specifically for the purpose). As much of the plant root system is included as possible.
	• In moving layers of soil, usually the top soil is removed and temporarily stored and then one or two subsequent layers of soil taken out down to a depth of 1 m. These are then replaced on the receptor sites in reverse order in an attempt to recreate the soil characteristics and growing conditions of the donor site.
	• Seeds can either be vacuumed up from the donor site or the whole site mown and the cuttings then collected. Seeds and cuttings are then distributed on the receptor site.
	• Habitat creation can involve translocation of species or habitats on to the site or, more typically, the introduction of commercially grown wild flower and grass seed, and the whips of young shrubs or trees. Some ground preparation will be necessary and in the case of wetland habitats may require significant earthmoving.
	• Species/habitat translocation and habitat creation are not precise arts and require expert advice from ecologists and landscape designers. The time of year and prevailing weather conditions can be crucial in terms of the success or otherwise of the operation, as can after care and management.
	• Note that it is illegal to dig up wild flowers without the landowner's consent (Wildlife and Countryside Act 1981) and in some cases species of plant and animal may require a licence before they can be moved.
	• In general animals move themselves although some species (particularly butterflies and other insects) may depend on a certain plant or habitat for their survival. Bats can be an exception to this as can some species of reptile or amphibian. Once again a licence may be needed before such species can be moved. The nests and eggs of birds are also protected under the Wildlife and Countryside Act 1981 and knowing destruction of a bird's nest is an offence. Certain species receive added protection.
	• Advice should be sought from English Nature (or its Scottish or Welsh equivalent) before attempting any habitat creation or translocation.
	• The clearance of areas of trees or scrub which may contain nesting birds is best avoided in the period April to July.

Issue C3.6.2	Habitat translocation and/or creation
Good practice references and further reading:	• Construction Industry Environmental Forum, *Nature conservation issues in building and construction*, Notes of a meeting held on 23/3/93, CIRIA, 1993. • Buckley, G.P. (Ed), *Biological Habitat Reconstruction*, Belhaven Press, London, 1989. • Baines, C. and Smart, J., *A Guide to Habitat Creation*, A London Ecology Unit Publication, Packard Publishing Ltd., Chichester, 1991. • Emery, M., *Promoting Nature in Towns and Cities: A Practical Guide*, Croom Helm, London, 1986.
Legal references:	See C2.1.10 on the provisions of the Wildlife and Countryside Act 1981.

Issue C3.6.3	Early planting of trees and shrubs
Background:	The early planting of trees and shrubs can help establish the character and form of the landscape setting of a development. By the time construction is complete the planting scheme has already had chance to settle down and to begin to mature. This not only helps soften the impact of the development but also demonstrates an early commitment to creating an appropriate landscape setting and can help market building developments by providing an attractive backdrop. It may also help provide continuity of wildlife habitats, either those lost to development or in helping to create a wildlife corridor across the site.
Background references:	See good practice references.
Good Practice:	• The extent of construction activities may mitigate against early planting. Any early planting will need protecting from construction activities (see C3.6.1). • The soil type at the site may also require early planting either because of longer than average periods being needed for the trees to establish themselves or because they will help prevent soil erosion after disturbance. • In creating wildlife habitats it is very often preferable to use small tree and shrub stock (eg. whips or saplings) which will grow in harmony with their surroundings. Because small stock is cheaper, larger quantities can be used and a greater failure of plantings tolerated. • In more formal landscapes larger trees and shrubs especially grown in nurseries can be transplanted. Special machines may be needed and great attention will have to be paid to after-care, particularly in the first 6 months or so. • Trees and shrubs stand a better chance of establishing themselves if planting is best carried out in late autumn or early spring. • Consider planting quick-growing plants to protect the main plants from windblow. • In certain situations a temporary planting scheme may be appropriate to provide seasonal colour and variety whilst the permanent landscape takes time to mature.
Good practice references and further reading:	• Baines, C., *Landscapes for New Housing; The Builder's Manual*. New Homes Marketing Board 1990. • Buckley, G.P. (Ed), *Biological Habitat Reconstruction*. Belhaven Press, London. 1989. • Construction Industry Environmental Forum, *Nature conservation issues in building and construction*, Notes of a meeting held on 23/3/93, CIRIA, 1993. • *More Homes and a Better Environment*: Report by the New Homes Environmental Group. This report focuses on the problems of providing affordable housing in an attractive environmental setting. In particular it assesses the role of the planning system and what makes good design. • *Trees on Development Sites*. Arboricultural Association. 1985.

Issue C3.6.4	Building structure and facades
Background: ▸ C3.6.1 ▸ D3.4.2 ▸ D2.3.12 ▸ D3.5.3	In certain situations a decision may have been taken to incorporate existing building structures or facades within a new development. Generally this will be for architectural, aesthetic or historical reasons and will have been identified as being necessary with the help and advice of the local authority, local community and special interest groups. Any archaeological remains will require similar protection as that afforded sensitive habitats.
Background references:	No specific references identified.
Good Practice:	• Any site preparation works including demolition should have left structures to be retained in a safe condition and readily identifiable. Frameworks should be erected to support facades and any other vulnerable structures. Party walls should be weather proofed. • All construction workers should be briefed as to the importance of such structures and the constraints put on them in terms of working practices, use of machinery etc. in their vicinity.
Good practice references and further reading:	• Goodchild, S.L. and Kaminski, M.P., Retention of major facades, in *The Structural Engineer*, Vol.67 No.8, 18 April 1989, pp 131–138. • Highfield, D., *The construction of new buildings behind historic facades*, E & F N Spon, London, 1991. • Thorburn, S., and Littlejohn, G.S., *Underpinning and retention*, Blackie Academic and Professional, 2nd Edn., Chapter 2, 1993.

C3.7 Traffic management

Issue C3.7.1	Legislation and policy
Current legal position:	The traffic management element of a development is likely to be dealt with by: • planning conditions, or • an agreement between the owner of the land and the developer on the one hand and the local planning authority or the highway authority on the other, the agreement being drawn up under either town and country planning or highways legislation. There are three main concerns for those authorities in highways and traffic terms:- • that there is sufficient capacity in the local road network to accommodate the additional traffic to be generated by the proposed development; • that there is safe access to and from the site both during and after construction; • that there is no adverse effect on general amenity as a result of the development. Planning permission may be refused if provision in any of these areas is considered unacceptable. In many cases conditions will be imposed on the planning permission to ensure the proposed development will be satisfactory; a local or highways authority may also require an applicant to enter into an agreement under Section 106 of the Town and Country Planning Act 1990 or Section 278 of the Highways Act 1980 before granting planning permission to deal with related matters which cannot be made the subject of a planning condition. *Planning Conditions* The conditions attached to a planning permission may deal with many aspects of traffic management, depending on the circumstances of each case. They may, for example, specify the means of vehicular access; require specified sight lines; prohibit occupancy until an acceptable means of vehicular access has been completed; provide for parking and loading/unloading spaces and internal circulation and layout; limit the points of access and hours of operation of construction vehicles and service vehicles.

Issue C3.7.1	Legislation and policy
Current legal position continued:	*Agreements* There are three kinds of agreement which may deal with traffic management: • **Section 106 Agreement** – Under Section 106 of the Town and Country Planning Act 1990, an agreement may be made between the landowner and developer and the local planning authority, to regulate the use or development of land either permanently or during construction. It is not limited to highway or traffic management matters. It is not unusual for a Section 106 agreement to address the same traffic management issues as planning conditions, but it can cover anything; a common use is to provide for improvement to the surrounding highway network either by carrying out specified works or making a payment to enable the local highway authority to do so. Section 106 was previously Section 52 of the Town and Country Planning Act 1971; many Section 52 agreements are still in force. • **Section 38 Agreement** – Section 38 of the Highways Act 1980 (as amended by the New Roads and Streetworks Act 1991) permits highway authorities to agree to adopt a proposed road provided it is built to certain specified standards. The authority will undertake to maintain the highway at public expense after construction and a maintenance period. It can address all traffic management issues. • **Section 278 Agreement** – Under Section 278 of the Highways Act 1980 (as amended by the New Roads and Streetworks Act 1991), an agreement for contributions towards or payment for highway works by persons deriving special benefit from them can be made. Any of the agreements may require a developer to find a traffic management order under the Road Traffic Act 1988 which is required in connection with the proposed development e.g. prohibiting vehicles over a certain weight from using a specified route. *The New Roads and Street Works Act 1991* This Act imposes an obligation on highway authorities (known as street authorities) to co-ordinate works on the highway in order to ensure safety, minimise inconvenience to street users and protect the structure of a street. Minimum notice periods are set according to the category of work to be carried out e.g. emergency works, minor works with excavation or major works. There are also requirements as to proper working practice (e.g. adequate lighting, guarding of works and the use of adequate signs) on everyone working within the highway.
References to the current legal position:	• Town and Country Planning Act 1990 (as amended by Planning and Compensation Act 1991). • Town and Country Planning (Scotland) Act 1972. • Highways Act 1980. • New Roads and Street Works Act 1991.
Policy and forthcoming legislation:	Guidance to local authorities on the use of conditions in planning permissions and model conditions for their use are contained in DoE circular 1/85. DoE circular 16/91 gives advice on the proper use of planning obligations under Section 106. There is continuing debate on what matters fall properly to be dealt with under either category; any planning conditions or requirements under agreements must be complied with even if they do not accord with general guidance pending action to remove or amend such restrictions. Codes of Practice under the New Roads and Street Works Act 1991 also deal with the practical implementation of that Act and supporting regulations.
Policy references:	• DoE Circular 1/85: *The Use of Conditions in Planning Permissions.* • DoE Circular 16/91: *Planning Obligations.* • DoE Planning Policy Guidance (PPG13): *Highway Considerations in Development Control*, 1990. • HAUC, *Code of Practice for the Co-ordination of Street Works and Works for Road Purposes and Related Matters*, 1992. • DoT, *Safety at Street Works and Road Works – A Code of Practice*, 1993. • DoE and DoT *Reducing Transport Emissions Through Planning*, 1993.

Issue C3.7.2	Traffic management strategies and their implementation
Background: ▸ D3.3.3 ▸ D3.3.4	Traffic generated by site operations can provide the largest impact of a civil engineering or building project on the surrounding area and communities and, as a result, needs active planning and management. Traffic management around roadworks, or works in or near the roadway which affect traffic flow, will need to take account of the New Roads and Street Works Act 1991 and its associated regulations and codes of practice.
Background references:	No specific references identified.
Good Practice: ▸ C2.1.2 ▸ C3.5.2	*In drawing up plans for managing traffic related to the site, ensure careful and practical consideration is given to the following:* • the extent to which the provisions of the New Roads and Street Works Act apply to the project and any necessary procedures to implement them; • the need as far as practicable to separate pedestrians and vehicles; • the need for traffic controls through lights or other means where site haul roads cross public roads; • the need to control dust, noise and vibration; • the desirability of preventing excessive exhaust emissions from encroaching on nearby properties; • avoiding the generation of congestion on public roads through excessive interference of site traffic with public traffic, including the provision for off-site queuing of delivery vehicles to prevent near-site congestion; • the possible improvement of local roads and/or junctions to accommodate site traffic; • the special problems of dealing with disposal of waste – see C2.1.2 and 3.5.2; • generally minimising hazards to others and the disturbance caused to neighbours and the local community.
Good practice references and further reading:	• Department of Transport, *Safety at Street Works and Road Works – A Code of Practice*, HMSO, 1992. • Highway Authorities and Utilities Committee, *Code of Practice for the co-ordination of Street Works and Works for Road Purposes and Related Matters*, HMSO, 1992.
Legal references:	• New Roads and Street Works Act 1991 and related regulations.

Issue C3.7.3	Production on- and off-site compared
Background:	One of the activities most likely to generate traffic to and from a site is the delivery of materials and components to be built into the works. For some of these, such as concrete production, there may be two alternative strategies to consider – production off-site and transport of the complete material to site for use when required, and production on site from more basic materials that can be brought to site in bulk at less frequent intervals. Available space for storage and site plant is obviously a major potential constraint.
Background references:	No specific references identified but see references in cross-referenced sections.
Good Practice:	*When planning construction projects:* • although it is recognised that most decisions in this area are almost always commercial, consider the environmental aspects of options of on- and off-site production of construction materials and include the environmental impact in the decision criteria;

Issue C3.7.3	Production on- and off-site compared

Good Practice continued:
 	
▸ C2.3.4
▸ C2.6.3
▸ C2.6.4
▸ C2.7.4
▸ C2.8.4
▸ C10.4
▸ D3.12
▸ D3.13
▸ D4.7
	• for example:
	  – how would each option influence the number of truck movements into and out of the site and their timing and frequency during the working day or week;
	  – which might create the greater dust nuisance or problems of control;
	  – what might be the energy consumption implications;
	• in particular, consider the comparative nuisance to neighbours and the local community of the two options;
	• for appropriately located projects, do not ignore the possibility of bringing bulk materials to the site or to close by by rail, river, canal or pipeline rather than road;
	• review relevant guidance elsewhere in the handbooks, in particular:
	  – purchasing – C2.3.4 and C6.3:
	  – materials – C6.4, C7.4, C8.4 and C10.4;
	  – design – D3.12, D3.13 and D4.7.

Good practice references and further reading:	No specific references identified but see references in cross-referenced sections.

C3.8 Environmental impact of temporary works

Issue C3.8	Environmental impact of temporary works
Background:	With so much concentration on site in meeting the technical specification for the permanent works and in meeting the project environmental requirements discussed earlier, it is easy to forget the potential environmental impact of temporary works designed and built solely by the contractor. They deserve consideration on environmental grounds in their own right. The guidance provided in the companion Handbook on Design and Specification will be relevant here, as indicated under good practice below.
Background references:	No specific references identified.

Good Practice:
 	
▸ D3.12
▸ D3.13
▸ D4.7
	In designing temporary works:
	
	• refer to the appropriate provisions of the companion to this volume – the Design and Specification Handbook and in particular:
	  – Section D3.12 – Criteria for primary material selection
	  – Section D3.13 – Use of materials
	  – Section D4.7 – Materials, in particular D4.7.5, sourcing of timber;
	• provide appropriate protection to the workforce and the public;
	• consider the visual impact of site hoardings on the neighbourhood and take active steps to reduce any adverse impact;
	• accept that there will be a conflict between safety provisions and aesthetics and that safety provisions are paramount here;
	• consider the ease of re-use when selecting materials for temporary works.

Good practice references and further reading:	• Somerville, S.H., *Control of groundwater for temporary works*, CIRIA Report 113, 1986. • Irvine, D.J, and Smith, R.J.H., *Trenching practice*, CIRIA Report 97, 1983, reprinted 1992. • Bielby, S.C., *Site Safety – A handbook for young construction professionals*, CIRIA Special Publication 90, 1992. • Venables, R.K et al, *Environmental Handbook for Building and Civil Engineering Projects, Volume 1: Design and specification*, CIRIA Special Publication 97, 1994.

C3.9   Environment in the site offices

Issue C3.9	Environment in the site offices
Background:	Although the main thrust of the guidance in this handbook is to improve the environmental performance of site operations, much can be done to reduce the environmental impact of construction sites by improving the environment in the site offices. There are five main areas for potential improvement.

Purchasing policies – The environmental performance of the site office should be considered at the purchasing or hiring stage and the guidance here used to take account of aspects such as insulation and other aspects of thermal performance, ventilation, re-usability etc.

Reducing energy costs and CO_2 emissions – Conserving fuel and energy is important in the construction process. Overall sums spent by construction firms on energy use on site are substantial, in excess of £1 billion a year. Whilst a good proportion of this results from use of plant on site it is estimated that by rationalising site lighting, heating and power, savings in the order of 20% could be realised with average savings of 10%.

Heating, lighting and ventilation – Efficiently heated and lit site offices save money for the contractor and should result in more productive use of time on the part of the site management workforce. Daylight and natural ventilation to site offices are desirable, both for user preference and to reduce energy use.

Water use – Water supplied to a site is essentially a clean and safe commodity. Water is a valuable resource and equally as important as its quality is the quantity used.

Waste management – Site staff generate waste paper, cardboard and packaging materials in their day-to-day work as in other offices. Bottles and cans are also waste products from catering on site. To encourage efficient collection and recycling of paper and other waste, the provision of adequate storage space and separate waste containers would help in sorting material prior to taking to a waste collection centre.

Noise – Noise generated during construction operations can have a tangible impact on the health and efficiency of site staff and site operatives. Unnecessarily high internal noise levels may be the result of poor planning and location of offices on site. |
| *Background references:* | • CIC Environment Task Group, *Our land for our children: an environmental policy for the construction professions*, Construction Industry Council, August 1992.
• CIEC Environment Task Force, *Construction and the Environment*, Building Employers Confederation, May 1992, Sections on Waste and Recycling, Energy Conservation and Efficiency, and Noise, BEC, 1992.
• Miller, S., *Going Green*, JT Design Build, Bristol.
• Laing Technology, *The Laing Environment – Environmental Policy Statement and practice notes*, 1990 onwards. |
| *Good practice references and further reading:* | • As background references. plus:
• Baldwin, R., Bartlett, P., Leach, S.J. and Attenborough, M., BREEAM 4/93, *An environmental assessment for existing office buildings*, BRE, 1993.
• *Energy Efficient Lighting for Buildings*, BRECSU-THERMIe, 1992.
• Contact BRECSU (Building Research Energy Conservation Support Unit) on 0923 664258. |

Issue C3.9	Environment in the site offices
Good Practice:	The following measures may seem comparatively trivial in scale compared to the cost and potential profits or losses involved in running major construction projects that need major site offices, but simple, straightforward measures requiring little time and/or effort can still result in significant cost savings as well as environmental benefits.

Site offices – lighting:

* choose cabins providing a window area of at least 10% of floor area and locate in unobstructed parts of the site to achieve good levels of daylight;
* use low energy fluorescent tubes for artificial lighting, and consider use of high pressure sodium lighting on construction projects running for over a year;
* switch off lights when leaving the offices, even for short periods.

Site offices – ventilation:

* locate site offices externally and ensure there are adequate openable windows for natural ventilation;
* if using part of an existing building for site offices ensure there are adequate openable windows for natural light and ventilation.

To minimise energy use:

* locate site offices in sheltered positions on site and fit valances around edges of raised offices to prevent air movement beneath them;
* use energy efficient, point-of-use heaters;
* identify and deal with potential sources of unnecessary ventilation heat loss by using, for instance, self-closers on doors and sealing gaps around service openings through external walls and floors;
* fit point-of-use water heaters and cookers in site kitchens;
* monitor energy use and take corrective action if abnormal consumption is noticed;
* locate and site power supplies and distribution systems to avoid excessive power line losses;
* consider wind turbines or photo-voltaic cells combined with rechargeable batteries for providing power to remote sites or to remote parts of extended sites such as road or pipeline projects.

In using water:

* consider low water flush toilet cisterns or water-less toilets, and low flush urinals, to minimise water consumption;
* install water meters to reduce consumption and provide early leak detection;
* consider installation of point-of-use water filters at sinks and drinking water outlets.

To minimise paper use:

* consider recycled paper wherever possible and limit copies to essential distribution;
* make provision for collection of scrap waste paper to re-use for drafts, messages etc.;
* use electronic mail and filing systems where practical.

To promote recycling:

* provide space within site offices for storage of waste paper and card;
* provide containers for recyclable waste such as glass bottles, steel and aluminium cans in a screened enclosure near to site offices.

To reduce noise levels in site offices:

* anticipate areas of noisy activity on site and locate site offices away from them wherever practicable, or screen offices to reduce potential noise problems;
* list noisy operations/plant likely to be employed during various phases of the project and consider their likely impact on site offices;
* ensure operatives are made aware of their responsibilities to minimise noise nuisance.

C3.10 Special considerations for specific civil engineering projects

Issue C3.10	Special considerations for specific civil engineering projects
Background:	Much of the guidance in this Handbook is generally applicable to all building and civil engineering projects, and some is more directed towards urban sites on which it is more likely that building projects will be under construction.
	With civil engineering projects, there are some special considerations that either only apply to civil engineering, or apply to individual types of project, for example dealing with dust and fumes blown from the ventilation systems for tunnelling projects.
	This sections presents brief guidance on these special considerations.
Background references:	• Construction Industry Environmental Forum, *Environmental issues in construction – A review of issues and initiatives relevant to the building, construction and related industries*, CIRIA Special Publications 93 and 94, 1993.
	• *Environmental Assessment: A guide to the identification, evaluation and mitigation of environmental issues in construction schemes*, CIRIA Special Publication 96, 1993.
	• CIC Environment Task Group, *Our land for our children: an environmental policy for the construction professions*, Construction Industry Council, August 1992.
	• CIEC Environment Task Force, *Construction and the Environment*, Building Employers' Confederation, May 1992.
	• Venables, R.K et al, *Environmental Handbook for Building and Civil Engineering Projects, Volume 1: Design and specification*, CIRIA Special Publication 97, 1994.
Good practice references and further reading:	• *Environmental Assessment: A guide to the identification, evaluation and mitigation of environmental issues in construction schemes*, CIRIA Special Publication 96, 1993.
	• Croners, *Environmental Management*, with quarterly amendment service, Croner Publications Ltd, First Edn, October 1991.
	• Skoyles, E.R. and Skoyles, J.R., *Waste Prevention on Site*, Mitchell Publishing, 1987.
	• Construction Industry Environmental Forum, *Recycling on site – the practicalities*, Notes of a meeting held on 22 June 1993, CIRIA, 1993.
	• Blake, L.S. (Ed), *Civil Engineer's Reference Book*, Butterworth.
	• *Timber: Types and sources*, Publication L296, Friends of the Earth, 1993.
	• *The Good Wood Manual*: Specifying Alternatives to Non-renewable Tropical Hardwoods, Friends of the Earth, January 1990.
	• Hall, K and Warm, P., *Greener Building Products and Services Directory*, Association for Environment Conscious Building Directory, Second Edition, 1993.
	• Planning Policy Guidance on Transport, PPG 13, Department of the Environment, April 1993.
	• DoE and DoT, *Reducing transport emissions through planning*, HMSO, April 1993.
	• Department of Transport, *The Good Roads Guide, Land form and alignment*, HA 55/92, HMSO, 1992.
	• Department of Transport, *The Good Roads Guide, New Roads: Planting, Vegetation and Soils*, HA 56/92, HMSO, 1992.
	• Department of Transport, *The Wild Flower Handbook*, HA 67/93, HMSO, 1993.
	• Edwards, A.C. and Mayhew, H.C., *Recycled asphalt wearing courses*, Report RR225, Transport Research Laboratory, 1989.
	• Cornelius, P.D.M and Edwards, A.C., *Assessment of the performance of off-site recycled bituminous material*, Report RR305, Transport Research Laboratory, 1991.
	• *Monitoring of cold road recycling process on a heavily-trafficked road*, ETSU New Practice Report NP/60, 1992.

Issue C3.10	Special considerations for specific civil engineering projects
Good Practice ▸ C3.7.2 ▸ C2.2.1	Use all the guidance in this handbook and the companion volume on design and specification that applies to your project and recognise that many civil engineering projects include a wide variety of building structures. For particular types of civil engineering project, use the following outline guidance in addition. *Roads and other pavement works* • The construction of new roads, particularly in rural areas, can be a very sensitive issue, as a current project in southern England demonstrates. Site management and the workforce need to be very aware that some local people may be very hostile to the project and this may even require additional security measures to protect the project and to protect potential protesters from risks on the site. • Road projects also tend to use comparatively large amounts of raw materials, so appropriate sourcing, and the transport considerations, need greater-than-average attention. • There are many opportunities for recycling road construction materials, either back into the re-constructed road if the specification so allows or into other road projects as fill, hardcore or sub-base. • Work on, in or next to roads will also come under the provisions of the New Roads and Street Works Act 1991, its associated regulations and related codes of practice such as Safety at Street Works and Road Works (1993). *Railway work* • Here, a major difficulty is often the need to work at night or on Sundays, when noise control will be a particular concern. • Access points to the work site may also be limited, so additional environmental disturbance may be involved some distance from the actual work. *Tunnelling projects* • These often involve 24-hour working, so the location of the site offices and the control of noise from above-ground operations need special care to avoid unacceptable nuisance during the night. • One significant potential concern is how to deal with spoil from the tunnel and how it is classified from the waste disposal point of view. • Another special problem may be dealing with the dust and fumes arising from the exhaust from the tunnel ventilation system – the fitting of filters, and their regular cleaning and maintenance may be necessary. *Dams and other water engineering projects* • These two are often consumers of large quantities of raw materials, but that issue is normally a major design consideration related to the project's location – the form of dam may be related to construction material availability. • They are often also in remote areas where concern for local landscape and/or wildlife habitats may be higher than usual. • Projects involving dredging may also produce large quantities of surplus material to be dealt with. *Renewable energy projects* • Here, there will almost certainly be heightened awareness of environmental issues, so that adherence to the environmental operational requirements of the contract may be even more important and under scrutiny than on other projects. Purchasing of materials and components to meet the environmental requirements of the specification are also likely to have very high priority. *Green purchasing policies* • Any overall purchasing policy guidance given in C2.2.1 may be difficult to apply to many civil engineering projects simply through a lack of available information. The AECB Guide and the *Good Wood Manual* are helpful but concentrate on building products and uses. Environmental criteria for assessing construction materials is being developed by CIRIA under a separate project (RP461).

Stage C4 Demolition and site clearance

Stage 4: Demolition and site clearance, it is assumed that most major demolition (to be the subject of a separate checklist) has already been dealt with but general demolition and site clearance are still to be completed. Remediation of contamination may already have been undertaken or may follow this stage.

C4.1   Legislation and policy

C4.2   Dealing with residual and/or unforeseen contamination in land and/or buildings and works

C4.3   Waste management

C4.4   Fires and their control

C4.5   Identification and protection of existing services

C4.6   Archaeology and ecology

C4.1 Legislation and policy

Issue C4.1	Legislation and policy
Current legal position:	Although demolition of works of buildings constitutes development, express planning permission is not required. Demolition of dwelling houses and buildings adjoining them has deemed planning consent under the General Development Order, provided that prior notification is given to the local planning authority so that it may determine whether approval will be required to the method of demolition or any proposed site restoration. Demolition of all other buildings is exempted from planning control (Town and Country Planning (Demolition) (No.2) Direction 1992). None of the above applies to partial demolition of any building. In those cases express planning permission is required. Separate notification of demolition to the local authority, electricity and gas suppliers, and adjacent occupiers, may be required under the Building Act 1984. Listed Building Consent and Conservation Area Consent will be required where relevant.
▶ C2.1.8	Planning control may also be relevant to site clearance. The methods used for site clearance e.g. the movement of waste out of the site and materials onto the site, may be the subject of planning conditions. Site clearance may also require the removal of contaminated land prior to the building stage. Breach of a planning condition may lead to enforcement action by the local planning authority.
▶ C2.1.2 ▶ C2.1.3	Any waste which is produced during the demolition and site clearance stages will render the developer subject to the duty of care as respects waste under the Environmental Protection Act 1990 (EPA). Also in respect of waste generated the waste management provisions of the EPA may apply where the developer keeps, treats or disposes of waste on site such as demolition waste.
▶ C2.1.4	Demolition and site clearance activities may pose a pollution risk to water courses through the treatment of contaminated soil. Under the Water Resources Act 1991 it is an offence to discharge poisons, toxins or polluting material into controlled waters. For discharges into controlled waters or the sewerage system discharge consents are needed under the water legislation. Particular attention should be paid to the identification, treatment or removal of hazardous materials either during the demolition of a building or during site clearance for example if contaminated land is involved, and asbestos will need to be removed by a specialised licensed contractor.
▶ C2.1.5	Noise pollution during demolition and site clearance will be governed by the relevant legal controls.
▶ C2.1.9 ▶ C2.1.6	Smoke from premises or dust from industrial trade or business premises may be a statutory nuisance where it is prejudicial to health or is a nuisance (Section 79 EPA). The local authority may issue an abatement notice on the person responsible or alternatively an aggrieved individual may apply to the Magistrates' Court for an order to abate the nuisance.
▶ C2.1.7	Demolition and clearance work conducted on site will be governed by the general duties imposed by the Health and Safety at Work etc Act 1974 and the numerous regulations made under it. Hazards may include flame cutting and exposure to toxic metal fumes.
References to the current legal position:	• Town and Country Planning Act 1990 or the Town and Country Planning (Scotland) Act 1972. • Planning and Compensation Act 1991. • Town and Country Planning (Demolition – Description of Buildings) (No.2) Direction 1992. • Town and Country Planning General Development Order 1988. • Building Act 1984. • Environmental Protection Act 1990. • Water Resources Act 1991. • Water Industry Act 1991. • Control of Pollution Act 1974. • Construction (General Provisions) Regulations 1961 (SI 1961 No. 1580, as amended). • Asbestos (Licensing) Regulations 1983 (SI 1983 No. 1649). • Control of Asbestos at Work Regulations.

Issue C4.1	Legislation and policy
Policy and forthcoming legislation:	No specific policy issues identified.
Policy references:	• DoE Planning Policy Guidance (PPG 18), *Enforcing Planning Control*, 1991. • DoE Circular 26/92, *Planning Controls over Demolition*. • DoE Circular 14/91, *Planning and Compensation Act 1991 – Guide to the Act*. • DoE Circular 21/91, *Planning and Compensation Act – Implementation of Main Enforcement Provisions*.

C4.2 Dealing with residual and/or unforeseen contamination in land and/or buildings and works

Issue C4.2	Dealing with residual and/or unforeseen contamination in land and/or buildings and works
Background:	However well a site investigation is done, and even if remedial treatment of contamination has been undertaken, some residual contamination is likely and there remains the possibility of unforeseen contamination being discovered during demolition and construction. Contractors and resident engineers and architects need to have in place contingency procedures for how to deal with such unforeseen contamination if it is discovered.
Background references:	• Leach, B.A., and Goodger, H.K., *Building on derelict land*, CIRIA Special Publication 78, 1991. • Curwell, S.R., March, C.G. and Venables, R.K., (Eds), *Buildings and Health, The Rosehaugh Guide to the Design, Construction, Use and Management of Buildings*, RIBA Publications, 1990, Chapter 9: Contaminated Land by Viney and Rees.
Good Practice:	• *At the demolition and site clearance stage*, contractors' project managers should have in place, prepared in consultation with the client and/or his professional advisors, plans for dealing with the otherwise unforeseen discovery of contamination or supposed contamination, for example, the unearthing of an unidentifiable drum which might contain toxic chemicals, or exposure of underground tanks or liquids in basements. • *After demolition*, consider undertaking a post-demolition survey to establish the actual levels and areas of any residual contamination to act as a basis for future action and development.
Good practice references and further reading:	• See background references, plus: • Construction Industry Environmental Forum, *Contaminated land*, Notes of a meeting held on 12/1/93, CIRIA, 1993. • Steeds, J.E., Shepherd, E. and Barry, D.L., *A guide to safe working practices for contaminated sites*, Unpublished CIRIA Core Programme Funders Report, FR/CP/9, July 1993, (in preparation as an open publication). • *Remedial treatment of contaminated land*, in 12 volumes, forthcoming CIRIA publications due to be published early 1994. • *Guidance on the sale and transfer of contaminated land*, Draft for open consultation, CIRIA, October 1993.
Legal references:	See legislation and policy issues C2.1.8 on Contaminated land, C2.1.2 on Waste and C2.1.3 on Duty of care and C2.1.4 on Water.

C4.3 Waste management

Issue C4.3.1	**Waste management principles at the demolition and site clearance stage**
Background: ▸ C2.1.2 ▸ C2.1.3 ▸ C2.3.8 ▸ C4.3.2	The current legal position on waste is set out in some detail in C2.1.2 and C2.1.3, and guidance on the development of waste management plans given in C2.3.8. It will be clear from that section that site management must take its duties and responsibilities for waste management seriously in order to comply with the legislation. This is particularly relevant at the demolition and site clearance stages since the nature of all the waste to be disposed of may be difficult to identify. There are five key areas which can be actively managed to ensure compliance and to effect sound environmental practice: • inspection of waste disposal contractors; • traffic management; • salvage and recycling; • dealing with asbestos and other known hazardous materials; • dealing with waste water, and with oil and petrol tanks. The first four are dealt with in this issue; the last is dealt with in C4.3.2 below.
Background references:	• Construction Industry Environmental Forum, *Recycling on site – the practicalities*, Notes of a meeting held on 22 June 1993, CIRIA, 1993. • Construction Industry Environmental Forum, *Recycling in construction: The use of recycled materials*, Notes of a meeting held on 13/7/93, CIRIA, 1993. • *Demolition and Reuse of Concrete and Masonry*, Proceedings of the 3rd International RILEM Symposium, October 1993. • Macneil, J., Wasted Opportunities, in *Building*, 19 March 1993. • Skoyles, E.R. and Skoyles, J.R., *Waste Prevention on Site*, Mitchell Publishing, 1987.

Good Practice:
 	
▸ C2.3.8
 	
▸ C3.7.1
▸ C3.7.2
	In implementing waste management plans at this stage:
	
	• review the guidance given in C2.3.8 and the plans that should have been drawn up at that stage:
	• in particular:
	  – make sure there is appropriate **inspection and verification of disposal contractors' licences** before they are engaged, and generally satisfy yourself that their operations match your own company standards in this area;
	  – ensure there are in place detailed procedures for the transfer of waste to registered carriers and that all who need to be are fully aware of those procedures;
	  – take particular care over **traffic management**, especially if contaminated soil and other debris is being transported – see C3.7.1 and 3.7.2;
	  – ensure there is active **salvage, recycling and sorting** of all appropriate materials such as bricks, concrete, blacktop, timber, window frames and tiles, classify site waste and separate it for recycling and, if not already identified, search locally for disposal outlets for recyclable materials;
	  – be alert to problems arising from waste disposal including residual paints and solvents in containers, dusts from cement, timber and asbestos, and broken glass all of which may cause safety hazards and/or pollution problems.

Good practice references and further reading:	• Building Employers' Confederation, *Business Bulletin enclosure on Disposal of Waste, Rubble and Demolition Waste*, 1992. • Building Employers' Confederation, *Consignment note for the carriage and disposal of Building Waste* (including Warnings and Advice, April 1992. • Building Employers' Confederation, *Technical Factfile Checklist on the Environmental Protection* (Duty of Care) Regulations 1991. • *Demolition Specification*, The National Federation of Demolition Contractors.
Legal references:	See C2.1.2 and C2.1.3.

Issue C4.3.2	Waste water control, and oil and petrol tanks
Background: ▸ C2.1.4 ▸ C3.5.1	Demolition sites always produce some waste water, sometimes in substantial quantities, as well as more obvious pollutants. For example, demolition activities can affect the water environment in many ways or produce many sources of waste water: • the run-off from washing-down of trucks and other equipment; • the hosing of dirt and waste from various surfaces; • leakage from oil and fuel tanks; • oil or fuel spillage through poor protection, vehicle damage or accidental valve opening; • vandalism; • the dumping of debris into or near to watercourses; • demolition of tanks without prior investigation and/or emptying. The disposal of these wastes needs to be carefully planned for and controlled. In addition, it is all-too-easy to hose away a spillage, say, of spilt oil or petrol without considering the consequences. At risk are the local rivers and other fresh water, the ground water and, in more urban areas, workers at work in drains and other sewerage facilities who can so easily be overcome with the fumes from hosed away chemicals. The National Rivers Authority is sympathetic to the problems faced by the construction industry at all stages and is available to assist demolition organisations with the provision of information, as well as providing the guidelines referred to below. However, the threat of and actual prosecution can be expected to increase as part of the NRA's pollution control policy, for anyone or any company that shows a disregard for the pollution dangers construction can pose.
Good Practice:	See C3.5.1 for guidance and references, and in particular the NRA Pollution Prevention Guidelines.

C4.4 Fires and their control

Issue C4.4	Fires and their control
Background:	Burning is often considered to be the only practical way of disposing of at least some debris from demolition and site clearance, but it is all too often the easy way out of the procedures for waste disposal. Some clients and contractors never burn demolition debris. Furthermore, the smoke, gases and fumes given off can cause significant pollution problems for the workforce and site neighbours.
Background references:	No specific references identified.
Good Practice:	*When clearing a site of final debris:* • consider carefully whether using a fire is the most environmentally appropriate action; • identify any bye-law restrictions on site fires; • if a fire is decided upon, make certain that the wind and other atmospheric conditions are not inappropriate, that it is kept under close control and that no potentially harmful or unknown substances such as unmarked chemicals drums are placed on it; • work to the appropriate provisions of BS6147; • keep a powerful hose available (and connected to a suitable supply) for dousing partially or completely in case of accidental flare-ups or accidental fuelling of the fire with unsuitable materials; • test the water supply pressure from time to time.
Good practice references and further reading:	• BS6147: 1980 *Code of Practice for Demolition*, British Standards Institution, Milton Keynes.

C4.5   Identification and protection of existing services

Issue C4.5	Identification and protection of existing services
Background:	Although more an occupational health and safety issue, there are a few potential environmental concerns in this area. Obviously, the careful **location of gas pipes** running through a site is vital for safety of the site and neighbourhood, rather than pollution, but other services if damaged may be more of environmental concern, for example: • **sewers**, which if damaged could pollute the ground water and/or local watercourses; • **electricity cables** which may contain PCBs in the insulation – even if the cable is not live, damage may release highly undesirable chemicals into the atmosphere or ground; • **water mains** if damaged could lead to contamination being introduced to drinking water; • **trade effluent drains** to storage tanks.
Background references:	No specific references identified.
Good Practice:	If any existing services cross the site, add concern about potential environmental mishaps to the standard safety procedures for dealing with them, particularly if they are to be exposed for excavation or construction of foundations.
Good practice references and further reading:	No specific references identified.

C4.6   Archaeology and ecology

Issue C4.6	Archaeology and ecology
Background: ▸ D2.3.12	Areas of ecological interest, and archaeological artifacts and remains, will have been identified during site investigations. Strategies will have been developed in consultation with interested parties as to whether or not they are to be retained within the development, to be moved to a safe location or to be destroyed. During the demolition and site clearance phase of the construction process it is therefore of vital importance that these strategies are carried out as agreed. Any deviation even in terms of timing of events can damage existing good relationships with the local community and may harm the marketability of the end product.
Background references:	No specific references identified.
Good Practice: ▸ C3.6	Section C3.6 discusses some of the appropriate strategies to be adopted to protect or translocate sensitive areas. If a site is to be destroyed, particularly with archaeological sites, it may be that third party specialists will want to carry out site investigations and/or excavations beforehand. The precise nature and timing of such investigations should be agreed in writing before work begins. Site management, engineers and all construction workers should be made aware of any strategies to be adopted. On-going consultation with third parties should be carried out on a regular basis throughout the period of demolition and site clearance.
Good practice references and further reading:	• *Code of practice on archaeology*, The British Property Federation and the Standing Conference of Archaeological Unit Managers.

Stage C5   Groundworks (including earthworks)

Stage C5: Groundworks (including earthworks) covers the particular environmental issues concerned with excavation and filling, whether of soft or hard soils or rock, for foundations, including piling, or to make earth structures such as embankments, bunds or cuttings.

C5.1   Materials and processes of potential environmental concern at this stage

C5.2   Legislation and policy

C5.3   Dealing with residual and/or unforeseen contamination in land and/or buildings and works

C5.4   Temporary storage of spoil, disposal of excess spoil and importing earthworks materials

C5.5   Hydrological, archaeological and ecological considerations

C5.6   Geotechnical processes and ground engineering

C5.1 Materials and processes of potential environmental concern at this stage

Issue C5.1	Materials and processes of potential environmental concern at this stage
Checklist of the main concerns: ► C4.2 ► C2.1.2 ► C2.1.3 ► C5.4 ► C5.5 ► C5.6 ► C2.3/C3.5/C3.6 ► C2.2/C8.4.3 ► C4.5	At the groundworks and earthworks stage, the principal materials and processes of environmental concern are: • **The potential discovery in the ground of unforeseen contamination, gases and natural hazards such as fissures and voids** – However good a site investigation or subsequent treatment of contamination may be, some of these features may escape detection – see C4.2 for further guidance. • **Temporary storage of spoil, disposal of excess spoil and/or the importing of fill materials for ground level raising** – Concerns include the clarification of when excavated material is defined as waste – all surplus material is defined as 'controlled waste', the related need to use licensed disposal contractors to cart away such material to licensed landfill sites, the use of borrow pits and storage areas on site and the assessment of the quality of the imported materials – see C2.1.2 and C2.1.3 for a summary of the legal position and C5.4 for guidance. • **The potential to disrupt the hydrology, hydrogeology, archaeology or ecology of the area** – see C5.5 for guidance. • **Problems associated with geotechnical processes or ground engineering** – Such techniques may involve blasting, with associated problems of noise, vibration, dust and flying rock fragments, handling special chemicals, disposing of water pumped from the excavation or well points, or the use of ground freezing – see C5.6 for guidance. • **Special problems of wind-blown dust, traffic management, mud on roads** – see C2.3, C3.5 and C3.6 for guidance. • **Timber for temporary excavation support** – see sub-section C2.2 on purchasing policies and apply the principles of C8.4.3 to the timber needed at this stage. • **Dealing with existing services** – see C4.5 and apply the guidance there appropriately to services in areas where earthworks must be undertaken.

C5.2 Legislation and policy

Issue C5.2	Legislation and policy
Current legal position: ▸ C2.1.2 ▸ C2.1.3 ▸ C2.1.5 ▸ C2.1.6/C2.1.7 ▸ C2.1.8 ▸ C2.1.4	The earthworks stage of construction involves the preparation of the site for construction. It may include the creation of structures out of earth e.g. bunds, the flattening of land and the removal of contaminated land. There are no environmental legal requirements relating specifically to the earthworks stage. However, general aspects of environmental legislation may apply. For example: • the keeping, treating or disposal of waste may require a waste management licence – – see C2.1.2; • any waste generated during these operations will be subject to the duty of care regarding waste – see C2.1.3; • noise and vibration during these operations may be subject to legal requirements on noise control – see C2.1.5; • health and safety requirements relating to employees and sub-contractors conducting the works – see C2.1.6 and C2.1.7; • legal issues relating to the removal of and liabilities in connection with contaminated land – see C2.1.8; • discharges into controlled waters or sewerage system – see C2.1.4.
References to the current legal position:	See references in Sections C2.1.1 to C2.1.8.
Policy and forthcoming legislation:	No specific policy issues identified.

C5.3 Dealing with residual and/or unforeseen contamination in land and/or buildings and works

Issue C5.3	Dealing with residual and/or unforeseen contamination
Background and Good Practice: ▸ C4.2	The background and principles of good practice of dealing with residual or unforeseen contamination at the groundworks stage are essentially identical to the demolition and site clearance stage, except that the risks, and the chances of discovery of unforeseen contamination should be lower. However, there is a danger that those involved at the groundworks stage will assume that all potential contamination has already been dealt with and therefore not be on the lookout for it. In those circumstances, the risks may be higher, so be on your guard.

C5.4   Temporary storage of spoil, disposal of excess spoil and importing earthworks materials

Issue C5.4	Temporary storage of spoil, disposal of excess spoil and importing earthworks materials
Background: ▸ C2.1.2 ▸ C2.1.3	On disposal of excess spoil, see C2.1.2 and 2.1.3 for the legal considerations. The prime concerns of site management on temporary storage of spoil, for example topsoil to be re-used, and the disposal of excess spoil are: • the definitions of the material as waste, controlled waste or other material; • to do what is necessary to comply with the law; • to do what is practicable and commercially sensible to minimise the environmental impact of the disposal arrangements. On the importing of fill material, consider whether it is 'waste' and whether the construction site will therefore have to be regarded as a licensed fill. PFA, which is much used as a fill material, is under consideration at EC level to be banned because of its metallic content.
Background references:	• Construction Industry Environmental Forum, *Environmental issues in construction – A review of issues and initiatives relevant to the building, construction and related industries*, CIRIA Special Publications 93 and 94, 1993, Chapter 6 on resources, waste and recycling. • CIC Environment Task Group, *Our land for our children: an environmental policy for the construction professions*, Construction Industry Council, August 1992. • CIEC Environment Task Force, *Construction and the Environment*, Building Employers Confederation, May 1992.
Good Practice:	• Review the legal position set out in C2.1.2 and C2.1.3 and recognise the present uncertainties. • Keep abreast of developments in the definition of waste for construction projects and the publication of final Waste Management Licensing Regulations. • Review and comply with the DoE Code of Practice. • When storing topsoil in bunds, avoid if at all possible bunds greater than 2 metres high in order to prevent anaerobic conditions developing, with consequent loss of soil fertility when it is placed in its new location. • Take active steps to deal with run-off and potential pollution from temporary storage bunds and other stored materials. • Try to minimise the width of haul roads and other easements and encourage one-way travel of tracked vehicles to minimise slewing and consequent degradation of the soil structure. • If at all possible, be flexible in final landscaping by adjusting plans to make use of available excess soil.
Good practice references and further reading:	• DoE, *Waste Management: The Duty of Care – A Code of Practice*, 1991. • Construction Industry Environmental Forum, *Waste management*, Notes of a meeting held on 6/4/93, CIRIA, 1993. • Building Employers' Confederation, *Business Bulletin enclosure on Disposal of Waste, Rubble and Demolition Waste*, 1992. • Building Employers' Confederation, *Consignment note for the carriage and disposal of Building Waste* (including Warnings and Advice), April 1992. • Building Employers' Confederation, *Technical Factfile Checklist on the Environmental Protection* (Duty of Care) Regulations 1991. • Edwards, A.C. and Mayhew, H.C., *Recycled asphalt wearing courses*, Report RR225, Transport Research Laboratory, 1989. • Cornelius, P.D.M and Edwards, A.C., *Assessment of the performance of off-site recycled bituminous material*, Report RR305, Transport Research Laboratory, 1991. • *Monitoring of cold road recycling process on a heavily-trafficked road*, ETSU New Practice Report NP/60, 1992.
Legal references:	See C2.1.2 and C2.1.3.

C5.5 Hydrological, archaeological and ecological considerations

Issue C5.5.1	Water courses and site hydrology
Background: ▸ C2.1.4	The potential effects of construction activity on water courses and other aspects of site hydrology may be of concern to the local authority, local community and in particular to the National Rivers Authority (NRA) from whom land drainage consent may be needed. The NRA was set up under the Water Act 1989 and is charged with managing and improving the quality of water resources and protecting them from pollution. Earthworks may: • cause pollution of water courses from run-off; • silt up or change the form of the bed or banks of a river; • adversely affect the surface drainage patterns or result in the infilling of a floodplain. Alternatively water may need to be abstracted or impounded for certain purposes. All of these activities will require some form of licence or consent from the NRA before work begins. The use of heavy machinery, particularly on wet areas of a site, may result in soil compaction and disruption of drainage patterns. In coastal areas or near to large rivers or lakes, substantial earthworks may be necessary to prevent the ingress of water to a construction site. This may result in, for example, increased water turbidity, loss of intertidal areas and habitats, and physical changes to neighbouring areas of coast or estuary.
Background references:	• Government White Paper: *This Common Inheritance – Britain's Environmental Strategy*, 1990, pp162–169. • Construction Industry Environmental Forum, *Environmental issues in construction – A review of issues and initiatives relevant to the building, construction and related industries*, CIRIA Special Publications 93 and 94, 1993, see section 8.3.3.1 Contamination of Water, pp. 161–162.
Good Practice:	• Close liaison with the NRA (or its equivalent authorities in Scotland and Northern Ireland) in respect of these issues is the only way forward. The NRA can prosecute companies or organisations which commit offences under the Water Resources Act 1991. • Run-off from the site should be directed through suitable traps and/or to settlement lagoons prior to any outflow into a watercourse. • Other relevant parties should also be kept informed of the implications of earthworks to water courses, site hydrology and associated wildlife habitats. • The site environmental manager should be responsible for ensuring all agreements are kept to, that the workforce are fully informed and that ongoing liaison with interested parties is maintained.
Good practice references and further reading:	• *Environmental Assessment: A guide to the identification, evaluation and mitigation of environmental issues in construction schemes*, CIRIA Special Publication 96, 1993, in particular section on River and Coastal Engineering. • National Rivers Authority, *Pollution Prevention Guidelines Working at Demolition and Construction Sites*, NRA, July 1992. • See C3.5.1 for details of other NRA Pollution Prevention Guidelines.
Legal references:	• The Water Resources Act 1991. • See C2.1.4 for a summary of the legal position.

Issue C5.5.2	Archaeology and ecology
Background:	Earthworks may affect ecological and archaeological aspects of the site, for example by influencing land drainage or coastal hydrology, causing direct damage to recognised sites, disturbing animals with vehicle movements or creating dust pollution. Earthworks may also form part of the landscaping plans for a site.
Background references:	• *Guidelines for minimising impact on site ecology*, The Royal Society for Nature Conservation, 1992. • British Property Federation, *Code of Practice on Archaeology*.
Good Practice: ► C3.6	• Any sensitive areas should have been clearly marked and fenced off or removed off site at an earlier stage. • Intended earthworks should be discussed with interested parties, for example, the NRA, local authority planners, English Nature staff (or Scottish or Welsh equivalents) and County archaeologists. Consultation with these groups and the local community should be maintained through the site environmental manager. • Any landscaping should be done under the supervision of expert landscape designers and/or managers. Operatives skilled in the use of heavy machinery to create appropriate land forms such as ponds, may also be needed.
Good practice references and further reading:	• *Environmental Assessment: A guide to the identification, evaluation and mitigation of environmental issues in construction schemes*, CIRIA Special Publication 96, 1993. • *Guidelines for minimising impact on site ecology*, The Royal Society for Nature Conservation, 1992. • Venables, R.K et al, *Environmental Handbook for Building and Civil Engineering Projects, Volume 1: Design and specification*, CIRIA Special Publication 97, 1994, in particular Sections D2.3.12 on Cultural features and D3.3.2 on Ecological value of the site.

C5.6   Geotechnical processes and ground engineering

Issue C5.6	Geotechnical processes and ground engineering
Background:	Some of the more specialist processes undertaken at the earthworks stage have special potential environmental concerns which need to be taken into account. The principal processes include: • rock excavation; • piling; • ground improvement (for example by vibro-replacement and dynamic compaction); • de-watering; • grouting; • ground freezing. The use of bentonite in diaphragm walling may also cause some environmental problems.
Background references:	• Wynne, C.P., *A review of bearing pile types*, PG1, CIRIA, 1988.

Good Practice:
 	
▸ C3.5.4
▸ C3.5.3
 	
▸ C3.5.4
▸ C3.5.2 & C5.4
 	
▸ C3.5.1
	At the earthworks stage of construction, make and implement plans to deal with the following special environmental concerns:
	
	• with *rock excavation:*
	  – noise and vibration effects – see C3.5.4;
	  – dust – see C3.5.3;
	
	• with *piling operations:*
	  – the noise and vibration effects – see C3.5.4;
	  – the disposal of spoil from drilled/augured piles – see C3.5.2 and C5.4;
	  – the disposal of concrete and steel waste from the trimming piles tops – see C3.5.2 and C5.4;
	
	• with *ground freezing:*
	  – the protection of workforce and site neighbours from the special equipment and materials used;
	
	• with *de-watering*, dealing with the silt-laden waste water – see C3.5.1;
	
	• with *bentonite*, avoiding accidental spillage and resulting blocking of local drains and/or contamination of local watercourses;
	
	• with *dredging*, be aware that highly contaminated or polluted material may unexpectedly be brought to the surface.

Good practice references and further reading:	• See Issues referred to in the above guidance, plus: • Fleming, W.G.K and Sliwinski, Z.J., *The use and influence of bentonite in bored pile construction*, PG3, CIRIA, 1977. • Skipp, B.O. and Hall, M.J., *Health and safety aspects of ground treatment materials*, CIRIA Report 95, 1982, reprinted 1992.

Stage C6 Foundations

Stage C6: Foundations covers the particular environmental issues associated with the construction of foundations, ie excluding the excavation for foundations (dealt with in Stage C5, but still working in the ground with formwork and structural materials.

C6.1   Materials and processes of potential environmental concern at this stage

C6.2   Legislation and policy

C6.3   Purchasing requirements, and sourcing and transport of materials

C6.4   Materials

C6.5   Dealing with residual and/or unforeseen contamination

C6.1 Materials and processes of potential environmental concern at this stage

Issue C6.1	Materials and processes of potential environmental concern at this stage
Introduction to the main concerns: ▸ C6.4.2\C7.4.2 ▸ C3.5.3 ▸ C3.5.4 ▸ C2.1.2/C2.1.3 ▸ C3.5.2 ▸ C4.5	At the foundations stage, the principal materials and processes of environmental concern are few in number: • **the potential for the re-use of timber** from formwork, trench sheeting etc – see C6.4.2 and C7.4.3; • **the potential for accidental spillage or wind-blown distribution of materials**, for example formwork sprayed chemicals and cement – see C3.5.3; • the **control of environmental noise** – see C3.5.4; • **dealing with waste** – see C2.1.2, C2.1.3 and C3.5.2; • **dealing with existing services** – see C4.5. Dealing with these concerns mainly involves applying the principles outlined earlier to the particular circumstances on your site.

C6.2 Legislation and policy

Issue C6.2	Legislation and policy
Current legal position: ▸ C2.1.5 ▸ C2.1.4 ▸ C2.1.6/C2.1.7	The sinking of foundations on a construction site may have the following legal implications. Three areas of legal requirements may be relevant: • **noise law**, both statute and common law should be considered since noise and vibrations generated during the foundation operations may affect both site workers and local inhabitants (C2.1.5 gives greater detail); • **water law** (see C2.1.4) may be relevant, particularly in relation to potential pollution of the water table; • **health and safety law** requirements in respect of workers on site – as outlined in more detail in C2.1.6 and C2.1.7.
References to the current legal position:	See issues referred to above.
Policy and forthcoming legislation:	No policy developments specifically affecting this stage have been identified.

C6.3 Purchasing requirements, and sourcing and transport of materials

Issue C6.3	Purchasing requirements, and sourcing and transport of materials
Background: ▸ C2.2.1	An overall green purchasing policy should have been established for the project. The task here is to apply it to the particular materials used at the foundations stage – formwork, blinding concrete, structural concrete, reinforcing steel etc.
Background references:	• *Environmental Labelling*, Introductory Leaflets, UK Ecolabelling Board, undated. • *The EC Scheme and how it will work*, Factsheet No.1, UK Ecolabelling Board, July 1992. • *The EC Ecolabelling Scheme*, Guidelines for Business, CBI Leaflet undated.

Good Practice:
 	
▸ C2.2.1
 	
▸ C2.3.4 & C6.3
▸ C6.4, C7.4, C8.4 & C10.4
▸ D3.12, D3.13 & D4.7
	When purchasing, sourcing and transporting materials for the foundations of a building or civil engineering project:
	
	• implement the overall green purchasing policy for the project – see C2.2.1;
	• review the specification carefully to identify particular environmental requirements for the materials and components in the foundations;
	• review relevant guidance elsewhere in the handbooks, in particular:
	  – purchasing – C2.3.4 and C6.3;
	  – materials – C6.4, C7.4, C8.4 and C10.4;
	  – design – D3.12, D3.13 and D4.7;
	• from the references quoted and your own company sources, compile a list of suitable suppliers;
	• obtain materials locally or from sources within the region wherever possible;
	• if at all possible, organise transport to minimise trips, for example by having a number of small loads delivered or collected in one round trip rather than in several trips;
	• without compromising payment terms, order materials sufficiently in advance to ensure efficient delivery.

Good practice references and further reading:	• Hall, K and Warm, P., *Greener Building Products and Services Directory*, Association for Environment Conscious Building Directory, Second Edition, 1993. • *Timber: Types and sources*, Publication L296, Friends of the Earth, 1993. • *The Good Wood Manual: Specifying Alternatives to Non-renewable Tropical Hardwoods*, Friends of the Earth, January 1990. • Elkington, J. and Hailes, J., *The Green Consumer Guide*, Guild Publishing, London, 1989. • *A guide to the safe use of chemicals in construction*, CIRIA Special Publication 16, 1981. • Ove Arup & Partners, *The Green Construction Handbook – A Manual for Clients and Construction Professionals*, JT Design Build, Bristol, 1993.

C6.4 Materials

Issue C6.4.1	Labelling schemes
Background:	There is an increasing need amongst clients, designers, specifiers and buyers for information about materials and products to allow environmentally friendly choices to be made. A product environmental labelling scheme called the 'Blue Angel' scheme has existed in Germany since 1978.
	Two significant developments relating to labelling of products with environmental information have recently occurred. Firstly the EC has been considering amendments to the Dangerous Substances Directive to label substances dangerous to the environment. Secondly the EC Regulation (880/92) governing an eco-labelling scheme has been adopted and taken effect. The purpose is to impose uniformity on 'environmentally friendly' claims manufacturers have begun to make for their products. Regulation is likely to be on consumer goods rather than commercial products. However paints and light bulbs have already been subject to pilot studies and work has been initiated on insulating materials.
	The UK Ecolabelling Board has been established to run the EC Ecolabelling Scheme in this country. The scheme is at the moment voluntary, but evidence suggests that possession of an ecolabel by a manufacturer gives a competitive edge. It is also possible that the scope of the scheme may be widened to service industries or become mandatory.
	Criteria for award of a label will vary from product to product but are likely to focus on energy economy, both in the production and use of the product. Lack of data is likely to be a problem for some time, and the compilation of environmental information and the rationalisation of the way it is expressed would be of help to contractors as well as designers and specifiers.
Background references:	• *Environmental issues in construction – A review of issues and initiatives relevant to the building, construction and related industries*, CIRIA Special Publications 93 and 94, 1993, Chapter 8 – Legislation and Policy Issues. • *Environmental Labelling*, Introductory Leaflets, UK Ecolabelling Board, (undated). • Organisation to contact: UK Ecolabelling Board: 071-820 1199.
Good Practice:	*As the eco-labelling schemes develop:* • look out for eco-labelling of products as the scheme is taken up by manufacturers; • where no eco-label is available, seek information where appropriate from materials and products manufacturers about embodied energy and CO_2 emissions in manufacture, impact of extraction, pollution associated with manufacture, sustainability of resource, recyclability and recycled content, emissions in use and problems from disposal after use.
Good practice references and further reading:	• *Environmental Labelling*, DTI, London, a DoE/DTI Quarterly Publication. • *The EC Scheme and how it will work*, Factsheet 1, UK Ecolabelling Board, July 1992. • *The EC Ecolabelling Scheme*, Guidelines for Business, CBI Leaflet undated.

Issue C6.4.2	Handling and storage of materials
Background:	Work at the foundations stages often involves much handling of materials and components by individual workers as well as by machine. Particular care is needed to avoid exposing the workforce to extra hazard and to avoid the release of contaminants to the atmosphere through careless handling.
Background references:	• Curwell, S.R., March, C.G. and Venables, R.K., (Eds), *Buildings and Health, The Rosehaugh Guide to the Design, Construction, Use and Management of Buildings*, RIBA Publications, 1990, especially Section B on materials & the building structure. • Bielby, S.C., *Site Safety – A handbook for young construction professionals*, CIRIA Special Publication 90, 1992.

Issue C6.4.2	Handling and storage of materials
Good Practice:	*In handling and storing materials for the foundation stage:* • ensure formwork oils are handled with care to minimise risk of skin irritation and/or accidental release to the atmosphere and water environment; • ensure the workforce are trained to understand hazard labelling on packaging; • ensure careful storage to maximise the potential for re-use, particularly of formwork; • minimise site stockpiling by phased delivery and/or 'just-in-time' purchasing and delivery.
Good practice references and further reading:	• Bielby, S. C., *Site Safety – A handbook for young construction professionals*, CIRIA Special Publication 90, 1992. • *A guide to the safe use of chemicals in construction*, CIRIA Special Publication 16, 1981. • *A Guide to the Control of Substances Hazardous to Health in Construction*, Report 125, CIRIA, 1993. • Ove Arup & Partners, *The Green Construction Handbook – A Manual for Clients and Construction Professionals*, JT Design Build, Bristol, 1993.

Issue C6.4.3	Recycling of materials
Background:	The primary opportunities for recycling of materials at this stage are: • the increased re-use of timber for formwork; • the collection of reinforcement off-cuts and their sale as scrap steel.
Background references:	• Construction Industry Environmental Forum, *Environmental issues in construction – A review of issues and initiatives relevant to the building, construction and related industries*, CIRIA SPs 93 and 94, 1993, Chapter 6 on Resources. • Construction Industry Environmental Forum, *Recycling on site – the practicalities*, Notes of a meeting held on 22 June 1993, CIRIA, 1993. • Mersky, R.L., *The American Experience with Recyling of specific materials*, Paper at WASTEMAN, Birmingham, March 1993. • MacNeil, J., Wasted Opportunities, in *Building*, 19 March 1993. • *Demolition and Reuse of Concrete and Masonry*, Proceedings of the 3rd International RILEM Symposium, October 1993.
Good Practice: ▸ C2.3.8 ▸ C3.4	*In seeking to increase the recycling of materials at this stage:* • implement the plans developed under C2.3.8; • encourage the workforce to take increased care of materials used, particularly formwork timber; • make sure all re-usable timber has all nails removed before storage or re-use; • set up separate collection of recyclable waste and identify outlets – see C3.4.
Good practice references and further reading:	• See C3.4, plus: • Construction Industry Environmental Forum, *Recycling on site – the practicalities*, Notes of a meeting held on 22 June 1993, CIRIA, 1993. • Construction Industry Environmental Forum, *Recycling in construction: The use of recycled materials*, Notes of a meeting held on 13/7/93, CIRIA, 1993. • Skoyles, E.R. and Skoyles, J.R., *Waste Prevention on Site*, Mitchell Publishing, 1987. • Venables, R.K et al, *Environmental Handbook for Building and Civil Engineering Projects, Volume 1: Design and specification*, CIRIA SP 97, 1994, sections D3.13.1 on waste, recyling and re-use of materials and D4.7.8 on use of recycled materials. • *Waste Recycling and Environment Directory*, Thomas Telford, London 1993.

Issue C6.4.4	Disposal of spare materials, waste and containers
Background:	See Issues: C2.12 – Legislation and policy on waste C2.1.3 – Duty of care for waste C2.1.7 – Control of substances hazardous to health C2.3.8 – Waste management, storage and re-use.
Background references:	• CIEC Environment Task Force, *Construction and the Environment*, Building Employers Confederation, May 1992.
Good Practice:	*To deal effectively with spare materials, waste and containers at this stage:* • introduce 'good housekeeping' measures to control waste of materials on site, for example by keeping a tidy site, proper storage of materials, educating operatives to minimise waste, setting waste reduction targets; • in particular, abide by the COSHH Regulations; • have clear procedures to identify and record waste arising from changes or errors in design and construction; • be alert to problems arising from waste disposal including residual oils in containers, dusts from cement and timber, and broken glass all of which can cause safety hazards if not pollution problems; • classify site waste and separate it for recycling at the point of use; • collect materials pallets for return to materials/products manufacturers.
Good practice references and further reading:	• See references in C2.3.8, plus: • *A Guide to the Control of Substances Hazardous to Health in Construction*, Report 125, CIRIA, 1993.
Legal references:	See C2.1.2, C2.1.3 and C2.1.7.

C6.5 Dealing with residual and/or unforeseen contamination

Issue C6.5	Dealing with residual and/or unforeseen contamination
Background and Good Practice: ▶ C4.2	The background and principles of good practice of dealing with residual or unforeseen contamination at the foundations stage are essentially identical to the demolition and site clearance and earthworks stages, except that the risks, and the chances of discovery of unforeseen contamination are even lower than at the groundworks stage since all excavation should by this stage have been completed. However, be on your guard.

Stage C7 Structural work for building or civil engineering

Stage C7: Structural work for building or civil engineering covers the particular environmental issues associated with the construction of the superstructure or frame of a building or civil engineering works, ie after completion of the foundations, when most concerns about contaminated land should be over, but before the specialist tasks of constructing the envelope, fitting out and commissioning tasks are undertaken.

C7.1 Materials and processes of potential environmental concern at this stage

C7.2 Legislation and policy

C7.3 Purchasing requirements, and sourcing and transport of materials

C7.4 Materials

C7.1 Materials and processes of potential environmental concern at this stage

Issue C7.1	Materials and processes of potential environmental concern at this stage
Checklist of the main concerns: ▸ C6.4.2/C7.4.3 ▸ C3.5.3 ▸ C3.5.4 ▸ C2.1.2/C2.1.3 ▸ C3.5.2	At the structural work stage, the principal materials and processes of environmental concern are also few in number: • **the potential for the re-use of timber** from formwork etc – see C6.4.2 and C7.4.3; • **the potential for accidental spillage or wind-blown distribution of materials,** for example formwork sprayed chemicals and cement dust from scabbling – see C3.5.3; • the **control of environmental noise** – see C3.5.4; • **dealing with waste** – see C2.1.2, C2.1.3 and C3.5.2. Dealing with these concerns mainly involves applying the principles outlined earlier to the particular circumstances on your site.

C7.2 Legislation and policy

Issue C7.2	Legislation and policy
Current legal position: ▸ C2.1.2/C2.1.3 ▸ C2.1.5 ▸ C2.1.6/C2.1.7 ▸ C2.1.4 ▸ C2.1.9	The following aspects of this stage may give rise to a need for permits or give rise to liabilities under environmental legislation. • Waste generated on site is subject to the legal requirements relating to the duty of care and waste management generally as outlined in C2.1.2 and C2.1.3. • Noise pollution may be a problem. The legal considerations are specified in C2.1.5. • Health and safety requirements imposed by the Heath and Safety at Work etc Act 1974 and regulations enacted under it will apply – see C2.1.6 and C2.1.7. • The production of waste water likely to cause pollution to controlled waters or discharges to those waters or the sewerage system are regulated under the water legislation – see C2.1.4. • Controls on air pollution including dust, smoke and grit – see C2.1.9.
References to the current legal position:	See references in issues listed above.
Policy and forthcoming legislation:	No policy issues specifically affecting this stage have been identified.

C7.3 Purchasing requirements, and sourcing and transport of materials

Issue C7.3	Purchasing requirements, and sourcing and transport of materials
Background: ▸ C2.2.1	An overall green purchasing policy should have been established for the project. The task here is to apply it to the particular materials and processes at this stage.
Background references:	• *Environmental Labelling*, Introductory Leaflets, UK Ecolabelling Board, undated. • *The EC Scheme and how it will work*, Factsheet No.1, UK Ecolabelling Board, July 1992. • *The EC Ecolabelling Scheme*, Guidelines for Business, CBI Leaflet undated.

Good Practice:
 	
▸ C2.2.1
 	
▸ C2.3.4 & C6.3
▸ C6.4, C7.4, C8.4 & C10.4
▸ D3.12, D3.13 & D4.7
	When purchasing, sourcing and transporting materials for the structural work associated with a civil engineering or building project:
	
	• implement the overall green purchasing policy for the project;
	• review the specification carefully to identify particular environmental requirements for the materials and components in the structure;
	• review relevant guidance elsewhere in the handbooks, in particular:
	  – purchasing – C2.3.4 and C6.3;
	  – materials – C6.4, C7.4, C8.4 and C10.4;
	  – design – D3.12, D3.13 and D4.7;
	• from the references quoted and your own company sources, compile a list of suitable suppliers;
	• obtain materials locally or from sources within the region wherever possible;
	• if at all possible, organise transport to minimise trips, for example by having a number of small loads delivered or collected in one round trip rather than in several trips;
	• without compromising payment terms, order materials sufficiently in advance to ensure efficient delivery.

Good practice references and further reading:	• Hall, K and Warm, P., *Greener Building Products and Services Directory*, Association for Environment Conscious Building Directory, Second Edition, 1993. • *Timber: Types and sources*, Publication L296, Friends of the Earth, 1993. • *The Good Wood Manual: Specifying Alternatives to Non-renewable Tropical Hardwoods*, Friends of the Earth, January 1990. • Elkington, J. and Hailes, J., *The Green Consumer Guide*, Guild Publishing, London, 1989. • *A guide to the safe use of chemicals in construction*, CIRIA Special Publication 16, 1981. • Ove Arup & Partners, *The Green Construction Handbook – A Manual for Clients and Construction Professionals*, JT Design Build, Bristol, 1993.

C7.4 Materials

Issue C7.4.1	Labelling schemes
Background and Good Practice:	See C6.4.1 and apply to the materials to be purchased for the superstructure stage.

Issue C7.4.2	Handling and storage of materials
Background and Good Practice:	See C6.4.2 and apply to the materials used for the superstructure stage.

Issue C7.4.3	Salvage and recycling of materials
Background:	There is much government and public pressure to increase the use of recycled materials, yet specifications, including those from some government departments, specifically preclude their use. Care and ingenuity are therefore required to increase their use while still satisfying the performance requirements of the specification.
Background references:	• Construction Industry Environmental Forum, *Recycling on site – the practicalities*, Notes of a meeting held on 22 June 1993, CIRIA, 1993. • Construction Industry Environmental Forum, *Recycling in construction: The use of recycled materials*, Notes of a meeting held on 13/7/93, CIRIA, 1993.
Good Practice:	*To increase the salvage of materials and use of recycled materials:* • check the specification for acceptability and negotiate acceptance if needs be; • when purchasing recycled materials, for example timber for formwork, make relevant technical checks, for example: – make sure timber has no woodworm or rot; – check bricks for signs of frost damage; • ensure there is enough material to finish the job as it may be impossible to match later; • if appropriate, consider use of products made from recycled ingredients (Listing given in AECB Greener Building Directory 4:34).
Good practice references and further reading:	• Hall, K and Warm, P., *Greener Building Products and Services Directory*, Association for Environment Conscious Building Directory, Second Edition, 1993. • Construction Industry Environmental Forum, *Recycling on site – the practicalities*, Notes of a meeting held on 22 June 1993, CIRIA, 1993. • Skoyles, E.R. and Skoyles, J.R., *Waste Prevention on Site*, Mitchell Publishing, 1987. • Venables, R.K et al, *Environmental Handbook for Building and Civil Engineering Projects, Volume 1: Design and specification*, CIRIA Special Publication 97, 1994, sections D3.13.1 on Waste, recycling and re-use of materials and D4.7.8 on Use of recycled materials.

Issue C7.4.4	Particular issues in the use of timber
Background:	Consideration of environmental factors in the production and specification of timber is becoming a key issue for the construction industry. The issue catches the public imagination more than most, given consumers' increasing interest in environmental matters in general and the sustainable production of materials in particular, yet some sections of the industry feel that the nature of this interest is misconceived.
	The environmentalists' main challenge is for the industry to accept that such damage affects the ecosystem as a whole, rather than just the local logging areas, and that what is required is a watertight certification process which will benefit both producers and consumers.
Background references:	• Construction Industry Environmental Forum, *Purchasing and specifying timber*, Notes of meeting held on 23/02/93, CIRIA, 1993. • *Timber: Types and sources*, Publication L296, Friends of the Earth, 1993. • World Wild Life Fund for Nature, *The inevitability of timber certification*, Discussion Paper, 1991. • The Forest Stewardship Council, A Discussion Paper, 1993. • Forests Forever, *A campaign for wood*, Timber Trades Federation.
Good Practice: ► C6.4.4/C2.3.8	*In purchasing timber for use in the structural elements of civil engineering or building projects:* • consult BRE and TRADA if in doubt about the sourcing of specified timber or to seek more environmentally acceptable alternatives; • consider obtaining timber from sources which have qualified for the Good Wood Seal of Approval (full listing given in Friends of the Earth's *Good Wood Manual*) or those listed in the AECB guide; • use timber joinery products manufactured from softwood wherever possible; • where tropical hardwood is specified, for example for jetties and fenders, consider negotiating a change to a temperate hardwood species; • recognise that, until a certification scheme for well-managed forests(or an equivalent scheme) is established, it will not be possible to be certain that *any* source is sustainably managed; • maintain, and re-use whenever possible, timber and timber boards used for site protection; • carry out timber preservative treatment off-site using pressure impregnation, with on-site treatment kept to a minimum; • dispose of surplus preservatives with great care – see C6.4.4 and C2.3.8.
Good practice references and further reading:	• Construction Industry Environmental Forum, *Purchasing and specifying timber*, Notes of meeting held on 23/02/93, CIRIA, 1993. • Construction Industry Environmental Forum, *The use of timber in construction*, Notes of a meeting held on 23/9/93, CIRIA, 1993. • *Timber: Types and sources*, Publication L296, Friends of the Earth, 1993. • *The Good Wood Manual: Specifying Alternatives to Non-renewable Tropical Hardwoods*, Friends of the Earth, January 1990. • Hall, K and Warm, P., *Greener Building Products and Services Directory*, Association for Environment Conscious Building Directory, Second Edition, 1993. • Building Research Establishment, *A Handbook of Softwoods*, SO 39, 1977. • Building Research Establishment, *Handbook of Hardwoods*, SO 7, 1972. • Building Research Establishment, *Remedial wood preservatives: use them safely*, Digest 371, 1992. • Building Research Establishment, *Wood-based panel products: their contribution to the conservation of forest resources*, Digest 375, 1992. • Contacts: BRE Timber Division: Tel: 0923 894040. TRADA: Tel: 0494 563091.

Issue C7.4.5	Disposal of spare materials, waste and containers
Background:	See Issues: C2.1.2 – Legislation and policy on waste C2.1.3 – Duty of care for waste C2.1.7 – Control of substances hazardous to health C2.3.8 – Waste management, storage and re-use.
Background references:	• CIEC Environment Task Force, *Construction and the Environment*, Building Employers Confederation, May 1992.
Good Practice:	*To deal effectively with spare materials, waste and containers at this stage:* • introduce 'good housekeeping' measures to control waste of materials on site, for example by keeping a tidy site, proper storage of materials, educating operatives to minimise waste, setting waste reduction targets; • have clear procedures to identify and record waste arising from changes or errors in design and construction; • be alert to problems arising from waste disposal including residual COSHH-controlled additives in containers and dusts from cement, which can cause safety hazards if not pollution problems; • classify site waste and separate it for recycling at the point of use; • collect materials pallets for return to materials/products manufacturers.
Good practice references and further reading:	• See references in C2.3.8, plus: • *A Guide to the Control of Substances Hazardous to Health in Construction*, Report 125, CIRIA, 1993.
Legal references:	See C2.1.2, C2.1.3 and C2.1.7.

Stage C8   Building envelope

Stage C8: Building envelope covers the particular environmental issues associated with the construction of the envelope of buildings, principally the external facings, cladding and the fixing of windows.

C8.1   Materials and processes of potential environmental concern at this stage

C8.2   Legislation and policy

C8.3   Purchasing requirements, and sourcing and transport of materials

C8.4   Materials

C8.1 Materials and processes of potential environmental concern at this stage

Issue C8.1	Materials and processes of potential environmental concern at this stage
Checklist of the main concerns: ► C8.3 ► C8.4.3/C8.4.6 ► C8.4.4 ► C8.4.5 ► C8.4.7 ► C2.1.9	At this stage, the principal environmental concerns are: • the **environmental performance of suppliers** of the cladding materials – see C8.3 for guidance on purchasing; • the **handling and storage of materials** and, in particular, the avoidance of hazardous materials that have in the past been used in building envelopes; • the **potential for the use of recycled materials and the recycling of timber on site** – see C8.4.3 and C8.4.6; • the **use of paints and solvents** and the inadvertent contamination of the internal environment being constructed – see C8.4.4; • the potential for the cladding to consume **scarce materials** when less environmentally damaging alternatives may meet the performance requirements of the specification – see C8.4.5; • the **exclusion of CFCs and halons** – see C8.4.7; • the need to control dust and fumes – see C2.1.9.
Background references and further reading:	• Curwell, S.R., March, C.G. and Venables, R.K., (Eds), *Buildings and Health, The Rosehaugh Guide to the Design, Construction, Use and Management of Buildings*, RIBA Publications, 1990, in particular Section B on Materials and the building structure and Section C on the Internal and external environments.

C8.2 Legislation and policy

Issue C8.2	Legislation and policy
Current legal position: ► C2.1.2/C2.1.3 ► C2.1.6/C2.1.7 ► C2.1.4 ► C2.1.9	The environmental requirements likely to apply to this stage will be: • waste legislation – waste derived from this stage, for example, used packaging materials or left over materials, will be waste and hence subject to the legal requirements outlined in C2.1.2 and C2.1.3; • health and safety legislation – see C2.1.6 and C2.1.7; • noise control legislation – see C2.1.4; • air quality legislation – see C2.1.9.
References to the current legal position:	See references in issues listed above.
Policy and forthcoming legislation:	No policy issues specifically affecting this stage have been identified.

C8.3 Purchasing requirements, and sourcing and transport of materials

Issue C8.3	Purchasing requirements, and sourcing and transport of materials
Background: ▸ C2.2.1 ▸ C8.1	An overall green purchasing policy should have been established for the project – see C2.2.1. The task here is to apply it to the particular materials used in a building envelope. See C8.1 above.
Background references:	See C7.3.
Good Practice:	See C7.3 and apply to the building envelope materials.
Good practice references and further reading:	See C7.3.

C8.4   Materials

Issue C8.4.1	Use of timber
Background and Good Practice:	See C7.4.4 and apply to the materials to be purchased for the building envelope stage.

Issue C8.4.2	Labelling schemes
Background: ▶ D3.8.1 ▶ D4.4.1	There is an increasing need amongst clients, designers, specifiers and buyers for information about materials and products to allow environmentally friendly choices to be made. A product environmental labelling scheme called the 'Blue Angel' scheme has existed in Germany since 1978 and there is a need for a similar scheme here. Two significant developments relating to labelling of products with environmental information have recently occurred. Firstly the EC has been considering amendments to the Dangerous Substances Directive to label substances dangerous to the environment. Secondly an EC Regulation governing an eco-labelling scheme has been agreed and has taken effect. The purpose is to impose uniformity on the 'environmentally friendly' claims manufacturers have begun to make for their products. Regulation is likely to be on consumer goods rather than commercial products. However paints and light bulbs have already been subject to pilot studies and work has been initiated on insulating materials. The UK Ecolabelling Board has been established to run the EC Ecolabelling Scheme in this country. The Ecolabelling Scheme is voluntary, but evidence suggests that possession of an ecolabel by a manufacturer gives a competitive edge. Criteria for award of a label will vary from product to product but are likely to focus on energy economy, both in the production and use of the product. Lack of data is likely to be a problem for some time, and the compilation of environmental information and the rationalisation of the way it is expressed would be of help to contractors as well as designers and specifiers.
Background references:	• Construction Industry Environmental Forum, *Environmental issues in construction – A review of issues and initiatives relevant to the building, construction and related industries*, CIRIA Special Publications 93 and 94, 1993. • *Environmental Labelling*, Introductory Leaflets, UK Ecolabelling Board, undated. • Organisation to contact: UK Ecolabelling Board: 071-820 1199.
Good Practice:	*As the eco-labelling schemes develop:* • look out for ecolabelling of products as the scheme is taken up by manufacturers; • where no ecolabel is available, seek information where appropriate from materials and products manufacturers about embodied energy and CO_2 emissions in manufacture, impact of extraction, pollution associated with manufacture, sustainability of resource, recyclability and recycled content, emissions in use and problems from disposal after use.
Good practice references and further reading:	• *Environmental Labelling*, DTI, London, a DoE/DTI Quarterly Publication. • The EC Ecolabelling Scheme: Guidelines for Business, CBI 1993. • Atkinson, C.J., and Butlin, R.N, *Ecolabelling of building materials and building products*, BRE Information Paper 11/93, BRE, 1993. • *The EC Scheme and how it will work*, Factsheet No.1, UK Ecolabelling Board, July 1992.

Issue C8.4.3	Handling and storage of materials
Background:	Work at this stage of construction often involves greater than average handling of materials and components by individual workers rather than machine. Particular care is therefore needed to avoid exposing the workforce to extra hazard and to avoid the release of contaminants to the atmosphere through careless handling. Needless waste will also be avoided.
Background references:	• Curwell, S.R., March, C.G. and Venables, R.K., (Eds), *Buildings and Health, The Rosehaugh Guide to the Design, Construction, Use and Management of Buildings,* RIBA Publications, 1990, Section B: Materials and the Building Structure.
Good Practice:	*With materials for the building envelope:* • avoid use of products containing asbestos, but if there is no alternative then use a suitably licensed contractor; • ensure mineral fibre boards and quilts are handled with care to minimise risk of skin irritation; • ensure the workforce are trained to understand hazard labelling on packaging; • take measures to deal with asbestos, lead pipe and sheet and other hazardous materials in refurbishing existing buildings, the steps including identification, measurement, assessment, sealing, removal and disposal, again using licensed subcontractors for any asbestos; • ensure practical and secure storage and handling to minimise waste and maximise the potential for recycling.
Good practice references and further reading:	• See background reference, plus: • *A Guide to the Control of Substances Hazardous to Health in Construction,* Report 125, CIRIA, 1993.
Legal references:	• Control of Substances Hazardous to Health Regulations 1988 (SI 1988 No. 1657) (as amended by SI 1990 No.2025, SI 1991 No.2431 and SI 1992 No.2382). • Classification, Packaging and Labelling of Dangerous Substances Regulations 1984, (SI 1984 No. 1244, as amended).

Issue C8.4.4	Use of paints and solvents
Background:	The principal potential hazards and environmental concerns here relate to: • the materials used; • the potential contamination of the internal environment from such processes as off-gassing; • the potential hazard to the workforce from solvent fumes when applying paints, resins and related materials; • the potential of solvents to pollute the water environment.
Background references:	• Curwell, S.R., March, C.G. and Venables, R.K., (Eds), *Buildings and Health, The Rosehaugh Guide to the Design, Construction, Use and Management of Buildings*, RIBA Publications, 1990.
Good Practice:	*When using paints and solvents on the building envelope:* • ask suppliers for COSHH statements and material safety data sheets or specifications for each product used, study them and amend working practices if needs be; • avoid sealant and glazing compound formulations using asbestos fibre as filler or lead as the drying agent; • use lead-free paint and primers, varnish and wood stain systems; • where synthetic paints are specified, consider negotiating a change to organic/natural paints for wood, metal and external masonry, although this may entail slightly higher costs to the client (Listing given in AECB Greener Building Directory); • develop procedures for the use of synthetic paints, which can give off toxic vapours, particularly when used in confined spaces.
Good practice references and further reading:	• Hall, K and Warm, P., *Greener Building Products and Services Directory*, Association for Environment Conscious Building Directory, Second Edition, 1993. • Curwell, S.R., March, C.G. and Venables, R.K., (Eds), *Buildings and Health, The Rosehaugh Guide to the Design, Construction, Use and Management of Buildings*, RIBA Publications, 1990: for paints: pp 127–8, 132, 76–77 and 138–9, and for solvents: pp66, 67, 70, 291 and 473–4. • *Solvent vapour hazards during painting with white-spirit-borne eggshell paints*, BRE Information Paper 3/92, March 1992. • *A Guide to the Control of Substances Hazardous to Health in Construction*, Report 125, CIRIA, 1993.

Issue C8.4.5	Use of scarce materials
Background:	Occasionally, specifiers will call for the use of a scarce material which an environmentally responsible contractor may prefer not to use, for example a scarce mineral for facing blocks. If contractors are to negotiate an acceptable alternative, information is the key, not only about availability and sourcing but technical performance comparable with the material specified.
Background references:	• Construction Industry Environmental Forum, *Purchasing and specifying timber*, Notes of meeting held on 23/02/93, CIRIA, 1993. • *Timber: Types and sources*, Publication L296, Friends of the Earth, 1993. • Croners, *Environmental Management*, with quarterly amendment service, Croner Publications Ltd, First Edn, October 1991. • The Environment Council, *Business and Environment Programme Handbook*, 1992.

Issue C8.4.5	Use of scarce materials

Good Practice:	*To minimise the use of scarce materials:*
	
	• if you know of a performance-equivalent alternative that is more environmentally acceptable, negotiate for it to replace what was originally specified;
	• if no alternatives are available:
	  – ensure sufficient material is ordered to finish the job but with minimum waste;
	  – arrange secure storage, possibly off-site, to prevent loss or damage.

Good practice references and further reading:	• Hall, K and Warm, P., *Greener Building Products and Services Directory*, Association for Environment Conscious Building Directory, Second Edition, 1993. • *The Good Wood Manual: Specifying Alternatives to Non-renewable Tropical Hardwoods*, Friends of the Earth, January 1990. • Ove Arup & Partners, *The Green Construction Handbook – A Manual for Clients and Construction Professionals*, JT Design Build, Bristol, 1993.

Issue C8.4.6	Use of recycled materials
Background:	There is much government and public pressure to increase the use of recycled materials, yet specifications, including those from some government departments, specifically preclude their use. Care and ingenuity are therefore required to increase their use while still satisfying the performance requirements of the specification.
Background references:	See good practice references.

Good Practice:	*To increase the use of recycled materials:*
	
	• check the specification for acceptability and negotiate acceptance if needs be;
	• when purchasing recycled materials, make relevant technical checks, for example:
	
	  – make sure timber has no woodworm or rot;
	  – check tiles for signs of frost damage;
	
	• ensure there is enough material to finish the job as it may be impossible to match later;
	• consider use of products made from recycled ingredients (See AECB Guide).

Good practice references and further reading:	• Hall, K and Warm, P., *Greener Building Products and Services Directory*, Association for Environment Conscious Building Directory, Second Edition, 1993. • Construction Industry Environmental Forum, *Recycling on site – the practicalities*, Notes of a meeting held on 22 June 1993, CIRIA, 1993. • Construction Industry Environmental Forum, *Recycling in construction: The use of recycled materials*, Notes of a meeting held on 13/7/93, CIRIA, 1993. • Venables, R.K et al, *Environmental Handbook for Building and Civil Engineering Projects, Volume 1: Design and specification*, CIRIA Special Publication 97, 1994, in particular D3.13.1 on use of recycled materials. • *Waste Recycling and Environment Directory*, Thomas Telford, London 1993.

Issue C8.4.7	Exclusion of CFCs and halons
Background: ▸ C4.7.4	It is now generally accepted that emissions of CFCs and other chlorine-containing substances, including HCFCs and halons, are causing depletion of the ozone layer and that production of these substances must be phased out as soon as is possible. The Montreal Protocol on substances that deplete the ozone layer, and an EC Regulation, require a complete phase-out of the production and supply of CFCs by 1996. CFCs were developed during the 1930s as safe alternatives to the then existing refrigerants, most of which are highly toxic and some flammable. Their high stability is a desirable property for a refrigerant but it means they can persist long enough in the atmosphere to be transported to the stratosphere. When CFC molecules reach the stratosphere, they are broken down by solar radiation. This releases chlorine free radicals which catalyse the breakdown of the ozone in the stratospheric ozone layer, an essential barrier which prevents harmful ultra-violet (UV) radiation from the sun reaching the earth's surface.
Background references: ▸ D4.7.4	• Building Research Establishment, *CFCs in Buildings*, BRE Digest 358, 1992. • Butler, D.J.G., *Guidance on the phase-out of CFCs for owners and operators of air conditioning systems*, Building Research Establishment Report PD25/93, 1993. • Curwell, S.R., Fox, R.C. and March, C.G., *Use of CFCs in Buildings*, Fernsheer Ltd, 1988 (out of print). • *The Montreal Protocol on substances that deplete the ozone layer*, Montreal 16/09/87, Foreign Office Command Paper, Treaty Series No.19, HMSO, 1990.
Good Practice:	*To ensure avoidance of CFCs and halons:* • review BRE Digest 358 and take appropriate action for your project; • avoid use of rigid urethane foams (RUFs) including polyurethane (PUR) and polyisocyanurate (PIR), extruded polystyrene (XPS) foams, extruded polyethylene (XPE) foams and phenolic foams for insulation; • use alternative insulants to those listed above including mineral fibre, expanded polystyrene (EPS), cellulose fibre, cellular glass (Full listing in BRE Digest 358 Table 5) and insulants made from recycled paper; • CFC-free RUF, XPS or phenolic foams may also be used as alternative insulants, but check their performance carefully as material thickness may need to be increased over the CFC blown product.
Good practice references and further reading:	• Building Research Establishment, *CFCs in Buildings*, BRE Digest 358, 1992. • Butler, D.J.G., *Guidance on the phase-out of CFCs for owners and operators of air conditioning systems*, Building Research Establishment Report, PD25/93, 1993. • Curwell, S.R., March, C.G. and Venables, R.K., (Eds), *Buildings and Health, The Rosehaugh Guide to the Design, Construction, Use and Management of Buildings*, RIBA Publications, 1990. • Department of Trade and Industry, *CFCs and Halons, Alternatives and the scope for recovery for recycling and destruction*, DTI, HMSO, 1990.
Legal references:	• EC Regulation 594/91 (as amended) on substances that deplete the ozone layer.

Issue C8.4.8	Disposal of spare materials, waste and containers
Background and Good Practice:	See C7.4.5 and apply to the materials, waste and containers to be disposed of at the building envelope stage.

Stage C9 Mechanical and electrical installations and their interface with civil and building work

Stage C9: Mechanical and electrical installations and their interface with civil and building work covers briefly the interface between the different disciplines and the special steps they need to take to ensure that their independent actions do not jeopardise the environmental performance of the project.

C9.1 Legislation and policy

C9.2 Co-ordination between structural, civil and service contractors, and use of the BSRIA environmental code of practice

C9.1   Legislation and policy

Issue C9.1	Legislation and policy
Current legal position: ▸ C2.1.2 & 3 ▸ C2.1.5 ▸ C2.1.6 & 7	Legal requirements at this stage will relate to: • the control of waste on land and responsibility for all waste produced – see C2.1.2 and C2.1.3; • the control of waste water – see C2.1.4; • the control of noise – see C2.1.5; • the health and safety legislation – see C2.1.6 and C2.1.7.
References to the current legal position:	See references in the issues listed above.
Policy and forthcoming legislation:	No policy issues specifically affecting this stage have been identified.

C9.2   Co-ordination between structural, civil and services contractors, and use of the BSRIA environmental code of practice

Issue C9.2	Co-ordination between structural, civil and services contractors, and use of the BSRIA environmental code of practice
Background: ► D3.12.6 ► C4.6	In most civil engineering and building projects, it is usual for the major civil engineering or building works to be let to different contractors than those undertaking the mechanical or electrical work or the building services installation. For the project environmental policy to be fully implemented, it is essential that the normal management interfaces between the disciplines also include consideration of environmental policy and implementation of that policy in a coordinated manner. BSRIA's work in this field provides invaluable help to building services designers and contractors and to other disciplines working with them. The BSRIA Code has been piloted during 1992/93 and is due to be published in its final form in early 1994.
Background references:	• Halliday, S.P., *Building Services and Environmental Issues – The Background*, BSRIA Interim Report, April 1992. • Halliday, S.P., *Environmental Code of Practice for Buildings and Their Services*, BSRIA, 1994. • Construction Industry Environmental Forum, *BSRIA's Environmental Code of Practice*, Notes of meeting held on 27/4/93, CIRIA, 1993. • Hejab, M. and Parsloe, C., *Space and weight allowances for building services plant*, Technical Note 9/92, BSRIA, 1993. • *Design Briefing Manual*, AG1/90, BSRIA, 1990.
Good Practice:	*Site management should, when dealing with interfaces between the main civil or building contractors and the services contractors:* • ensure all parties are aware of the project's environmental policy and of each other's corporate environmental policies; • set up regular liaison between each of the contractors' environmental managers such that they ensure no conflict in environmental terms between the various contractors' actions; • ensure all contractors' environmental managers are aware of the strategy and guidance outlined in BSRIA's Environmental Code of Practice for Buildings and Their Services in assisting contractors to implement coordinated and practical, environmentally responsible procedures; • ensure environmental plans drawn up by the individual contractors following the guidance in this Handbook are complementary; • consider the preparation and distribution of contract 'trees' showing all the contractual and environmental inter-relationships on the project; • consider the implementation of punitive measures against any contractors and sub-contractors for non-compliance with the project environmental policy.
Good practice references and further reading:	• See background references. • Contacts: BSRIA (Building Services Research and Information Association) 0344 426511. BRECSU (Building Research Energy Conservation Support Unit) 0923 664258. CIBSE (Chartered Institution of Building Services Engineers) 081-675 5211.

Stage C10 Trades: joinery, painting, plastering etc.

Stage C10: Trades: joinery, painting, plastering etc covers the particular environmental concerns at the fitting out stage of buildings and the trades stages of civil engineering projects.

C10.1   Materials and processes of potential environmental concern at this stage

C10.2   Legislation and policy

C10.3   Purchasing requirements, and sourcing and transport of materials

C10.4   Use, handling and storage of materials and components by the finishing trades

C10.5   Commissioning

C10.1 Materials and processes of potential environmental concern at this stage

Issue C10.1	Materials and processes of potential environmental concern at this stage
Checklist of the main concerns: ▸ C10.3 – C10.4 ▸ C8.4.3/C8.4.6 ▸ C8.4.4 ▸ C8.4.5 ▸ C8.4.7 ▸ C2.1.9	At this stage, the principal environmental concerns are: • the **environmental performance of suppliers** of the materials involved – see C10.3 for guidance on purchasing; • **the handling and storage of materials** and, in particular, paints, varnishes, solvents, adhesives and sealants – see C10.4 • the **potential for the use of recycled materials and the recycling of timber on site** – see C8.4.3 and C8.4.6; • the **use of paints and solvents** and the inadvertent contamination of the internal environment being constructed – see C8.4.4; • the potential for the finishings to consume **scarce materials** when less environmentally damaging alternatives may meet the performance requirements of the specification – see C8.4.5; • the **exclusion of CFCs and halons** – see C8.4.7; • the need to **control dust and fumes** – see C2.1.9.
Background references and further reading:	• Curwell, S.R., March, C.G. and Venables, R.K., (Eds), *Buildings and Health, The Rosehaugh Guide to the Design, Construction, Use and Management of Buildings*, RIBA Publications, 1990, in particular Sections B on Materials and the Building Structure and Section C on the Internal and External Environments.

C10.2 Legislation and policy

Issue C10.2	Legislation and policy
Current legal position: ▸ C2.1.2 – C2.1.7	The most important legal issues likely to arise at this stage will be in relation to the disposal of leftover, off-cut materials, packaging and waste generally. The following are the most important and need active consideration. • The health and safety legislation should be considered in detail in view of potential irritants and toxic substances present in furnishing components and decorating materials eg volatile organic compounds in adhesives and wood preservatives and paint – see C2.1.6 and C2.1.7. • The control of waste on land and responsibility for waste produced is important – see C2.1.2 and C2.1.3. • The control of noise is covered under C2.1.4. • The control of waste water e.g. from plastering or painting is covered under C2.1.5. • The Environment Protection (Controls on Injurious Substances) Regulations 1992 prohibit the use of paint in which lead carbonate or lead sulphate is a constituent. There is an exception where such paint is to be used for a listed building or scheduled monument.
References to the current legal position:	**EC** • Directive relating to restrictions on the marketing and use of certain dangerous substances 76/769/EEC (as amended). **UK** • Environmental Protection Act 1990. • Environmental Protection (Controls on Injurious Substances) Regulations 1992 (SI 1992 No. 31).
Policy and forthcoming legislation:	For policy and forthcoming legislation on issues mentioned above, see the sections referred to. The Environmental Protection (Control on Injurious Substances) (No.4) Regulations are due to be enacted soon. They will restrict the use of cadmium in pigments used in plastic products and urea formaldehyde, and the use of cadmium in paints containing zinc.
Policy references:	• DoE Circular 3/92, *Environmental Protection (Control of Injurious Substances) Regulations 1992.*

C10.3 Purchasing requirements, and sourcing and transport of materials

Issue C10.3	Purchasing requirements, and sourcing and transport of materials
Background: ▸ C2.2.1	An overall green purchasing policy should have been established for the project. The task here is to apply it to the particular materials used by trades in fitting out a building or works.
Background references:	• See C7.3, plus: • *Environmental Labelling*, Introductory Leaflets, UK Ecolabelling Board, undated. • *The EC Scheme and how it will work*, Factsheet No.1, UK Ecolabelling Board, July 1992. • *The EC Ecolabelling Scheme*, Guidelines for Business, CBI Leaflet undated.
Good Practice:	See C7.3 and apply the guidance there to materials for trades activities.
Good practice references and further reading:	• See C7.3, plus: • Hall, K and Warm, P., *Greener Building Products and Services Directory*, Association for Environment Conscious Building Directory, Second Edition, 1993. • *Timber: Types and sources*, Publication L296, Friends of the Earth, 1993. • *The Good Wood Manual: Specifying Alternatives to Non-renewable Tropical Hardwoods*, Friends of the Earth, January 1990. • Atkinson, C.J., and Butlin, R.N, *Ecolabelling of building materials and building products*, BRE Information Paper 11/93, BRE, 1993. • Elkington, J. and Hailes, J., *The Green Consumer Guide*, Guild Publishing, London, 1989. • *Selection of air-to-air heat recovery systems*, TN 11/86, BSRIA, 1986

C10.4 Use, handling and storage of materials and components by the finishing trades

Issue C10.4	Use, handling and storage of materials and components by the finishing trades
Background:	The principles of environmentally responsible use, handling and storage of materials have been outlined previously in Sections C7.4.4, C7.4.5 and C8.4.2–7 above. The materials and processes of potential environmental concern at this stage are listed in C10.2 above. The aim here must be to effectively apply the principles to this stage.
Background references:	See C7.4.4, C7.4.5 and C8.4.2–7.
Good Practice: ▸ C7.4.4 ▸ C7.4.5 ▸ C8.4.2 – C8.4.7 ▸ C10.2	*There are eight aspects of materials use and handling to be covered:* • **Labelling schemes** • **Handling materials** • **Use of timber** • **Use of paints and solvents giving rise to volatile organic compounds** • **Use of scarce materials** • **Use of recycled materials** • **Exclusion of CFCs and halons** • **Disposal of spare materials, waste and containers.** The principles of dealing with these aspects are described in C7.4.4, C7.4.5 and C8.4.2 to C8.4.7. To achieve good practice at this stage, implement the guidance in those sections to the materials and processes described in C10.2. In addition, apply the following special considerations: • the use of adhesives, paints, varnishes and sealants which will not leave long-lasting, difficult-to-remove pollutants in the internal environment; • the use of lead-free paints; • the potential for dust from woodworking and fumes from pipe-jointing.
Good practice references and further reading:	See C7.4.4, C7.4.5 and C8.4.2 – 7.

C10.5 Commissioning

Issue C10.5	Commissioning
Background:	In the same way that mechanical and electrical plant need to be commissioned before a building or work of civil engineering is occupied or used in earnest, much of the trades fitting out needs to be checked and confirmed as functioning as intended. In addition, there may be a need to 'bake out' the furnishings and paint finishes before occupation to ensure low concentrations of volatile organic compounds that may off-gas from paints, varnishes, wall coverings and other furnishings.
Background references:	• Curwell, S.R., March, C.G. and Venables, R.K., (Eds), *Buildings and Health, The Rosehaugh Guide to the Design, Construction, Use and Management of Buildings*, RIBA Publications, 1990, Chapters C1: Indoor air quality, and C5: Building related sickness. • CIBSE Commissioning Codes, Chartered Institution of Building Services Engineers.
Good Practice:	*At the completion of the fitting out and trades stage:* • check for residual smells and other signs of airborne solvents, chemical vapours and gases, and ensure adequate ventilation or positive flushing out to disperse them before occupation of the building.
Good practice references and further reading:	• Halliday, S.P., *Environmental Code of Practice for Buildings and Their Services*, BSRIA, 1994. • *Pre-commissioning cleaning of water systems*, BSRIA, AG8/91. • *The Commissioning of Water Systems in Buildings*, BSRIA, AG2/91. • *The Commissioning of VAV Systems in Buildings*, BSRIA, AG1/91. • *The Commissioning of Air Systems in Buildings*, BSRIA, AG3/91. • CIBSE, *Commissioning Codes: A: Air distribution Systems (1971); B: Boiler Plant (1975); C: Automatic Controls (1973); R: Refrigerating systems (1991); W: Water distribution systems (1989)*, Chartered Institution of Building Services Engineers, London.

Stage C11 Landscaping, reinstatement and habitat restoration or creation

Stage C11: Landscaping, reinstatement and habitat restoration or creation covers the particular environmental issues associated with the landscaping of a site after completion of the main works, reinstatement of any particularly sensitive areas, and the restoration or creation of special wildlife habitats.

C11.1   Legislation and policy

C11.2   Purchasing requirements

C11.3   Planting and maintenance

C11.4   Landscape establishment, and habitat restoration or creation

C11.1 Legislation and policy

Issue C11.1	Legislation and policy
Current legal position: ▸ C2.1.2 – C2.1.5 ▸ C5.2	Once the building operations have been completed and the building fitted out, at this stage there are two main areas of legal requirements arising. • First, the project's planning permission and any waste management licence granted is very likely to have conditions attached to it. It is usual for those conditions to address the post-construction landscaping and reinstatement issues and it is a legal requirement in each individual case to comply with any relevant conditions attached to such permissions. Under the Town and Country Planning Act 1990 a breach of condition notice may be issued by the local planning authority where there is a breach of a planning condition. There is no right of appeal against the service of such a notice and there are criminal sanctions for non-compliance with the notice. There are also other enforcement powers under the town and country planning legislation. Under the Environmental Protection Act 1990 it is an offence to breach any condition of a waste management licence. • Secondly, legal requirements arise whilst physically conducting the landscaping, reinstatement and habitat restoration. These processes are likely to produce waste materials. Therefore the duty of care as respects waste (see C2.1.3) and the legislation on waste management generally (see C2.1.2) will apply. The legislation on noise C2.1.5) may be relevant in respect of the level of noise generated during these activities. Discharges to controlled waters or the sewerage system are governed by the issues outlined in C2.1.4. • The legal requirements for groundworks may also apply.
References to the current legal position:	• Town and Country Planning Act 1990. • Town and Country Planning (Scotland) Act 1972. • Planning and Compensation Act 1991. • Environmental Protection Act 1990.
Policy and forthcoming legislation:	No policy issues specifically affecting this stage have been identified.
Policy references:	• DoE Planning Policy Guidance (PPG 18), *Enforcing Planning Control*, 1991. • DoE Circular 14/91, *Planning and Compensation Act 1991 – Guide to the Act.* • DoE Circular 21/91, *Planning and Compensation Act – Implementation of Main Enforcement Provisions.*

C11.2 Purchasing requirements

Issue C11.2	Purchasing requirements
Background:	The restoration or creation of landscape and wildlife habitats as part of a development has varying impacts on the environment according to the materials used. Materials which are sourced locally will have lower embodied energy. The use of certain materials may have a detrimental effect on unsustainable resources far removed from the development site. For example, the use of peat by the horticultural trade is now threatening some of the country's best bog and wetland wildlife habitats. Tropical hardwoods are sometimes used in hard landscaping as are inappropriate mineral products. Certain fertilisers and pesticides may be harmful to wildlife and possibly even people.
Background references:	See good practice references.
Good Practice: ▸ C2.2.1 – C2.2.4	*In establishing purchasing policy specifically for landscaping:* • review the overall purchasing policy guidance given in C2.2.1–4; • where possible ensure materials used are of local provenance because local variations within species can be important in creating the correct wildlife habitats for that place and in retaining local distinctiveness, and local stone will look more at home than minerals imported from abroad or even another area of Britain; • avoid the unnecessary use of tropical hardwoods; • avoid the use of peat, or growing media containing peat, as alternatives are now readily available; • use plant species and numbers of an appropriate size for their setting and take any necessary measures to protect the planting whilst it becomes established; • use organic means of fertilisation and pest control where possible and especially in wildlife areas; • if it is necessary to use chemicals, use them in strict accordance with the manufacturer's instructions and any regulations in force; • consult the MAFF list of approved pesticides; • avoid using atrazine and simazine which are two weedkillers frequently found in drinking water at levels which breach EC standards.
Good practice references and further reading:	• Baines, C: *Landscapes for New Housing; The Builder's Manual*. New Homes Marketing Board 1990. • *The Environmental Charter for Local Government – Practical Recommendations*. Friends of the Earth, London, 1989. • Johnston, J. and Newton J., *Building Green – A Guide to using plants on roofs, walls and pavements*, London Ecology Unit, 1993. • *A guide to the safe use of chemicals in construction*, CIRIA Special Publication 16, 1981. • *A Guide to the Control of Substances Hazardous to Health in Construction*, Report 125, CIRIA, 1993. • Contact the British Association of Landscape Industries on: 0535 606139.

C11.3 Planting and maintenance

Issue C11.3.1	**Ground preparation**
Background:	Good ground preparation and planting techniques are essential for satisfactory growth, minimal losses, good establishment and ease of future maintenance.
Background references:	• See good practice references, plus: • Clouston, B. (Ed), *Landscape Design with Plants*, Heinemann, 1977. • *The planting and aftercare of trees and shrubs*, Countryside Commission. • Weddle, A.E., *Landscape Techniques*, Heinemann, 2nd Ed., 1979.

Good Practice:	*In preparing the ground for landscaping:*
	
	• The topsoil and subsoil should be stripped and stored separately, with mounds not exceeding 2 metres in height.
	• Check soil pH and chemistry and their compatibility with chosen plants.
	• In areas of new planting:
	  – substantial quantities of topsoil are needed to grow trees and shrubs, but only limited quantities are required to establish grassland;
	  – the contractor must secure clear guidance on soil-handling constraints;
	  – the formation on which soil is spread must be roughened to allow it to key in properly;
	  – any weed growth should be suppressed either by the use of appropriate chemicals or by mechanical or smothering techniques;
	  – larger specimens of trees will benefit from having their roots dipped in a moisture retentive material such as alginure and the introduction of slow release fertilisers may also be beneficial;
	  – planting trees in tree shelters and mulching the soil surface to suppress weed growth and increase moisture retention will also help them to become established;
	  – avoid use of peat-based materials to enrich soils;
	  – on-going management, particularly using hand tools, is crucial to the success of wildlife habitats.

Good practice references and further reading:	• *Landscape Design Guide: Volume 1 Soft Landscape*, PSA Projects. • *The Good Roads Guide, New Roads – Planting, Vegetation and Soils*, Department of Transport. • Hibberd, B.G., *Urban Forestry in Practice*. Forestry Commission Handbook 5, HMSO, London, 1989. • BSI, *Recommendations for general landscape operations*, BS4428, 1979. • Harte, J.D.C., *Landscape, Land Use and the Law*, E & F Spon, 1985.

Issue C11.3.2	Dealing with residual contamination
Background: ▸ D2.3.1 ▸ C2.5	All major contamination should have been identified and dealt with in an appropriate manner before reaching this stage of the construction process (see section D2.3.1 in the Design and Specification Handbook and C2.5 in this Handbook). However, small pockets of contamination may still remain to be discovered during preparation of the ground for planting or contamination may have occurred during the construction process. Examples of the latter could include excessive liming of soils around concrete batching plants, or pollution from oil spills and leakages. Contamination may affect the success of any planting scheme as well as providing an environmental hazard.
Background references:	• Venables, R.K et al, *Environmental Handbook for Building and Civil Engineering Projects, Volume 1: Design and specification*, CIRIA Special Publication 97, 1994, Section D2.3.1.
Good Practice:	*To deal with residual contamination at this stage:* • in most cases the most appropriate strategy for dealing with pockets of contamination would be to remove the contaminated soil to a licensed landfill site; • an alternative strategy may be to transfer the material to a part of the site designated for encapsulation or some other form of treatment; • clean material of similar characteristics in terms of pH, structure and composition should be used to replace the excavated soil. DoE research concerning guidance on plant species tolerant of contaminated ground was current in June 1993 and a report will be published in due course.
Good practice references and further reading:	• Baines, C: *Landscapes for New Housing; The Builders Manual.* New Homes Marketing Board 1990, (page 11). • Bradshaw, A.D. and Chadwick, M.J., *The Restoration of Land: The ecology and reclamation of derelict and degraded land*, Blackwell Scientific, Oxford, 1980. • Clouston, B. (Ed), *Landscape Design with Plants*, Heinemann, 1977. • Weddle, A.E., *Landscape Techniques*, Heinemann, 2nd Ed., 1979. • Cairney, T., *Reclaiming Contaminated Land*, Blackie, Glasgow and London, 1987.

C11.4 Landscape establishment, and habitat restoration or creation

Issue C11.4	Landscape establishment, and habitat restoration or creation
Background:	A good landscape can make a tremendous difference to the visual appearance and wildlife interest of a development or civil engineering project. Those people who use or live near a particular development or works will have an ongoing interest in such factors long after construction has been completed. Landscapes which have been created and/or restored to a high standard can enhance a company's environmental performance and reputation. Alternatively, landscaping failures can be expensive and unattractive and, particularly if relating to wildlife habitat creation or restoration, may dishonour obligations made to planning authorities, statutory organisations or local wildlife groups. The NRA has statutory duties in relation to the conservation and enhancement of the aquatic environment.
Background references:	See good practice references.

Good Practice:
 	
▸ C11.2
▸ C11.3.2
 	
▸ C3.6.4
	When establishing landscaping and/or restoring of creating habitats:
	
	• skilled operatives will be needed to carry out any ground modelling;
	• the successful establishment of plant material will depend on several factors:
	   – condition of material when purchased
	   – care taken in transportation and storage
	   – species choice appropriate to local conditions
	   – preparation of area to be planted
	   – appropriate sowing or planting methods
	   – after care and management;
	• trees and shrubs should have all minor damage repaired, should receive judicious pruning (according to location and species) and be planted at the correct depth and at the right season, ideally during mid-November to end March;
	• support or anchor trees where necessary to prevent wind blow;
	• use guards to protect young trees from damage by animals such as rabbits and deer;
	• sow seeds, either by hand or by machine, in early autumn or early spring, raking and lightly rolling the soil surface to firm the seedbed and assist establishment;
	• turfs should be transplanted on to especially prepared sites;
	• when creating and restoring wildlife habitats, particularly wetlands, pay careful attention to factors such as species choice, habitat design and the quantity and quality of water supply;
	• when restoring habitats, care must be taken not to damage or disturb existing features of interest whether plant or animal;
	• carrying out work at the right time of year can be crucial;
	• management and after-care will be crucial to ensure the success of any planting scheme, the exact nature of such work depending on the species, habitats or landscape involved – expert advice should be sought.

Good practice references and further reading:	• Baines, C: *Landscapes for New Housing; The Builders Manual.* New Homes Marketing Board 1990. • Baines, C., and Smart, J: *A Guide to Habitat Creation.* A London Ecology Unit Publication. Packard Publishing Ltd., Chichester, 1991. • Beckett, K and G: *Planting Native Trees and Shrubs.* Jarrold, Norwich. 1979. • BS4428: 1979 *Recommendations for general landscape operation*, BSI. • Buckley, G.P. (Ed) *Biological Habitat Reconstruction*, Belhaven Press, 1989. • Emery, M: *Promoting Nature in Towns and Cities: A Practical Guide*, Croom Helm, London, 1986. • Coppin, N.J. and Richards, I.G., *Use of vegetation in civil engineering*, CIRIA Book 10, 1990. • Lovejoy and Partners, *Landscape Handbook*, E & F N Spon. • Committee for Plant Supply and Establishment, *Plant Handling Code*, Horticultural Trades Association, Reading. • Committee for Plant Supply and Establishment, *Good Planting Guide*, Horticultural Trades Association, Reading.

Stage C12 Site reinstatement, removal of site offices and final clear away

Stage C12: Site reinstatement, removal of site offices and final clear away covers the particular environmental issues associated with the clearing of the site, reinstatement of any areas not covered at Stage 11, and the removal and disposal or preparation for re-use of the site offices.

C12.1 Legislation and policy

C12.2 Options for salvage and recycling

C12.3 Disposal of waste

C12.4 Reinstatement of the site

C12.1 Legislation and policy

Issue C12.1	Legislation and policy
Current legal position: ▸ C11.2 ▸ C2.1.2 – C2.1.3	As with Stage C11, site reinstatement is likely to be the subject of conditions attached to the planning consent and/or any waste management licence which has been granted. Section C11.2 deals with the effect of non-compliance with such conditions. Other legal requirements will relate to materials left at the end of the construction process ie. waste. These will be subject to the duty of care as outlined in C2.1.3 and other general waste requirements outlined in C2.1.2.
▸ C2.1.4	Discharge consents may be required and need to be complied with if water is discharged to controlled waters or the sewerage system.
▸ C2.1.5	Legal requirements relating to noise pollution control may be relevant if the process is noisy or causes a lot of vibration.
References to the current legal position:	See references in sections listed above.
Policy and forthcoming legislation:	No policy issues specifically affecting this stage have been identified.

C12.2 Options for salvage and recycling

Issue C12.2	Options for salvage and recycling
Background:	It is at this stage that a great many temporary facilities set up to support the construction process need to be dismantled. It is all too easy, in the rush to clear the site, to dispose of as waste materials and whole facilities such as temporary offices that could, with financial as well as environmental advantage be salvaged for re-used or recycling.
Background references:	• Construction Industry Environmental Forum, *Recycling on site – the practicalities*, Notes of a meeting held on 22 June 1993, CIRIA, 1993. • Laing Technology, *The Laing Environment – Environmental Policy Statement and practice notes*, 1990 onwards, in particular a practice note on re-use of site offices. • Skoyles, E.R. and Skoyles, J.R., *Waste Prevention on Site*, Mitchell Publishing, 1987.
Good Practice:	*At this late stage in the construction process:* • ensure the staff and workforce are aware of company policy on salvage, re-use and recycling of site facilities; • develop and implement the plan for site run-down based on a practical combination of financial and environmental objectives by: – returning to the main company plant and equipment yard all materials and facilities that can, even with some refurbishment, be re-used on future sites; – separating out, stock-piling by category and finding outlets for any other materials and used facilities that you cannot re-use but which could be recycled by others; – only disposing as waste debris and other materials which cannot economically be dealt with in the first two categories.
Good practice references and further reading:	• Contact University of Surrey Demolition and Recycling Centre, Tel: 0473 300800 Ext.2216.

C12.3 Disposal of waste

Issue C12.3	Disposal of waste
Background: ▸ C12.2 ▸ C2.1.2 ▸ C2.1.3 ▸ C2.3.8	This stage can generate a wide variety of wastes which need care in disposal – for example, general mixed debris, old fencing, parts of office buildings that cannot be re-used, spare part-filled chemical drums, and fuel storage. It may also be necessary to dispose of fuel storage protection bunds and oil and/or silt traps. As indicated in C12.2 above, it should be only the materials and facilities that cannot be re-used or recycled, but there may still be considerable volume involved. The legal position is summarised in: • C2.1.2 – Legislation and policy on waste, and • C2.1.3 – Duty of care; Other background information is provided in C2.3.8 on waste management.
Background references:	See references in sections listed above.
Good Practice: ▸ C2.3.8 ▸ C12.2	*In clearing the site:* • implement the plans developed in C2.3.8; • ensure that, as indicated in C12.2, only the debris and materials that cannot be re-used or recycled are disposed of as waste; • since at this stage many of the site management team may have moved on to new projects, ensure those remaining are aware of the legal position on waste and in particular the need to use licensed waste disposal contractors; • take particular care with the problems of mixed waste that is often generated at this stage – creation of large quantities of 'special waste' by inadvertent contamination of other waste by mixing with one container of a specialist chemical could prove very costly as well as environmentally unsatisfactory.
Good practice references and further reading:	See C2.1.2, C2.1.3 and C2.3.8.
Legal references:	See C2.1.2 and 2.1.3.

C12.4 Reinstatement of the site

Issue C12.4	Reinstatement of the site
Background:	Although the main landscaping has been dealt with at Stage C11, much tidying and re-shaping will need to be done once all offices and other facilities have been removed.
Background references:	See Section C11.
Good Practice:	Refer to the guidance given in Section C11 in particular C11.4 on landscaping and habitat restoration and creation.

Construction Stage 12: Site reinstatement, removal of site offices and final clear away

Stage C13 Handover, and guidance on maintenance and records

Stage C13: Handover, and guidance on maintenance and records covers the recommended actions to be undertaken at handover of buildings or civil engineering works to their owners and/or occupiers, including the maintenance recommendations and records that should be provided.

C13.1   Legislation and policy

C13.2   Records of materials used and treatment of contaminated land, and advice to owners and occupiers

13.1 Legislation and policy

Issue C13.1	Legislation and policy

Current legal position:	At the time of handover, developers and contractors will want to ensure so far as possible, that they will have no residual environmental liability once they have completed the project and left the site. There are several issues which should be considered from a legal point of view:
	
	• contractual provisions, for example, if the property is to be leased, the lease covenants must include provisions that the tenant will not cause pollution or harm to the environment, that the tenant will be responsible for waste and must seek the required environmental permits/consents;
	
	• for any environmental permits, consents, licences or authorisations which were granted to a developer/contractor in respect of the site:
	
	   – the developer/contractor should ensure that all conditions have been fully complied with;
	   – it should be determined of whether such consents should be transferred to tenants/occupiers – the legislation often provides for the transfer of permits on the sale or transfer of property, so the relevant legislation should be checked, the method of transfer identified and the procedure concluded.

References to the current legal position:	No specific references identified.
Policy and forthcoming legislation:	The draft Construction (Design and Management) Regulations impose duties in respect of health and safety at all stages from project formulation and development through to the post–construction phase. Two proposed requirements relate to the completion of the construction phase: • the handing over of a 'health and safety' file for each structure. Information in the file should include 'as built' design and details of maintenance facilities in order that these features can be taken into account if construction work is carried out at a later date; • if the building is leased the 'health and safety' file should be passed on to the leaseholder.
Policy references:	• Health & Safety Commission, *Proposals for Construction (Design and Management) Regulations and Approved Code of Practice*, 1992.

C13.2 Records of materials used and treatment of contaminated land, and advice to owners and occupiers

Issue C13.2	Records of materials used and treatment of contaminated land, and advice to owners and occupiers
Background:	Whatever the provisions that are finally brought in with the forthcoming Construction (Design and Management) Regulations (duc in mid-1994) and the associated Approved Code of Practice, it is increasingly regarded as good practice to provide owners and occupiers with records of the design principles, construction processes, and the key materials and components used, and with guidance on future maintenance procedures.
Background references:	• Health and Safety Commission, *Proposals for Construction (Design and Management) Regulations and Approved Code of Practice*, 1992.

Good Practice:	*When handing over a work of civil engineering or a completed building:*
	
	• the professional advisors and/or contractors should, in addition to any technical or contractual records of the facility, provide owners and occupiers' facilities managers with records of:
	
	  – any investigation undertaken to search for contamination of the site prior to construction starting;
	  – the treatment provided to remove or biologically reduce the contamination;
	  – the results of any post-treatment surveys;
	  – records of any potentially hazardous materials built into the project which may become exposed during regular maintenance, major overhauls or on demolition;
	  – guidance on the timing and extent of necessary maintenance of the works or building and of the equipment incorporated within it;
	
	• any notifications to the local authorities, police and fire brigade about changes in responsibilities;
	• services, plant and controls, together with maintenance schedules;
	• the project teams should stress the need for such records to be passed to future owners and occupiers' facilities managers;
	• any guidance handed to owners and occupiers should include advice on the landscaping, trees and shrubs planted and any special habitat that has been restored or created;
	• the project team should provide a guidance report to owners and occupiers setting out the initial environmental policy for the project, any variations to it agreed during the design or construction phases, and any guidance they believe will be helpful to the owners or occupiers in operating the works or building in an environmentally responsible manner;
	• the professional advisors should establish a record set of all such handover information.

Good practice references and further reading:	• Miller, S., *Going Green*, JT Design Build, Bristol. • *Design Briefing Manual*, AG1/90, BSRIA, 1990. • BS7750: 1992 *Specification for Environmental Management Systems*, British Standards Institution, Milton Keynes. • Construction Industry Environmental Forum, *Environmental Management in the Construction Industry*, Notes of meeting held on 22/09/92, CIRIA, 1992. • CBI, *Corporate environmental policy statements*, Confederation of British Industry, London, June 1992. • Barwise, J., and Battersby, S., *Environmental Training*, Croner Publication, 1993. • Halliday, S.P., *Environmental Code of Practice for Buildings and Their Services*, BSRIA, 1994. • *Operating and maintenance manuals for building services installations*, BSRIA AG1, 1987. • *Maintenance Management for Building Services*, CIBSE, TM 17, 1990. • *Building Services Maintenance*, BSRIA, RG No.1, Vol 1 and 2, 1990. • *Standard Maintenance Specification for Mechanical Services in Buildings*, HVCA. • *Decision in Maintenance*, BSRIA, TN14/92, 1992.

List of main background and good practice references

A Books, reports and guidance documents.

A Guide to the Control of Substances Hazardous to Health in Construction, Report 125, CIRIA, 1993.

Advisory Committee on Business and the Environment, *A guide to environmental best practice for company transport*, DoE, November 1992.

Atkinson, C.J., and Butlin, R.N, *Ecolabelling of building materials and building products*, BRE Information Paper 11/93, BRE, 1993.

Baldwin, R., Bartlett, P., Leach, S.J. and Attenborough, M., BREEAM 4/93, *An environmental assessment for existing office buildings*, BRE, 1993.

Barwise, J., and Battersby, S., *Environmental Training*, Croner Publication, 1993.

Bright, K., *Building a greener future – Environmental issues facing the construction industry*, CIOB Occasional Paper 49, Chartered Institute of Building, 1991.

BS7750: 1992 *Specification for Environmental Management Systems*, BSI, Milton Keynes.

BSI, *BS7750 2nd Edition Draft for Public Comment*, DC400220/93, BSI, London, 1993.

Building Research Establishment, *CFCs in Buildings*, BRE Digest 358, 1992.

Butler, D., and Howard, P.N., Life Cycle CO_2 Emissions: From the Cradle to the Grave, *in Building Services*, 1992.

Butler, D. J. G., *Guidance on the Phase-out of CFCs for owners and operators of Air Conditioning Systems*, PD25/93, Building Research Establishment, 1993.

CBI, *Corporate environmental policy statements*, Confederation of British Industry, London, June 1992.

CIBSE, *Building Energy Code*, in 4 parts, CIBSE, 1975–82.

CIC Environment Task Group, *Our land for our children: an environmental policy for the construction professions*, Construction Industry Council, August 1992.

CIEC Environment Task Force, *Construction and the Environment*, BEC, May 1992.

Climate and Site Development, BRE Digest No. 350, in 3 parts, BRE, 1990.

Construction Industry Environmental Forum, Notes of meetings:.

>   *Materials and product blacklists*, 23/7/92, CIRIA, 1992.
>   *BSRIA's Environmental Code of Practice*, 27/4/93, CIRIA, 1993.
>   *Building lifetimes*, 27/7/92, CIRIA, 1992.
>   *Environmental Management in the Construction Industry*, 22/09/92, CIRIA, 1992.
>   *Environmental considerations in public sector refurbishment*, 12/10/92, CIRIA, 1992.
>   *Water pollution from construction sites*, 20/10/92, CIRIA, 1992.
>   *Buildings and energy*, 25/11/92, CIRIA, 1992.
>   *Green Clients – The role of the client in setting the environmental tone of construction projects*, 30/11/92, CIRIA, 1992.
>   *Green buildings: the designer's perspective*, 8/12/92, CIRIA, 1993.
>   *Contaminated land*, 12/1/93, CIRIA, 1993.
>   *Ecolabelling: the implications for the construction industry*, 26/1/93, CIRIA, 1993.
>   *Purchasing and specifying timber*, 23/02/93, CIRIA, 1993.
>   *Considerate Builders and Contractors*, 9/3/93, CIRIA, 1993.
>   *The Environment, Economics and the Construction Industry*, 19/03/92, CIRIA, 1992.
>   *Nature Conservation Issues in Building and Construction*, 23/3/93, CIRIA, 1993.
>   *Waste Management*, 6/4/93, CIRIA, 1993.
>   *Life-cycle eco-analysis in building and construction*, 25/5/93, CIRIA, 1993.
>   *Recycling on site – the practicalities*, 22/6/93, CIRIA, 1993.
>   *Recycling in construction: The use of recycled materials*, 13/7/93, CIRIA, 1993.
>   *Contaminated land – Insurance and liabilities in sale and transfer*, 14/9/93, CIRIA, 1993.
>   *Contaminated land – Technologies and implementation*, 14/9/93, CIRIA, 1993.
>   *The use of timber in construction*, 23/9/93, CIRIA, 1993.

Construction Industry Environmental Forum, *Environmental issues in construction – A review of issues and initiatives relevant to the building, construction and related industries*, CIRIA Special Publications 93 and 94, 1993.

Coppin, N.J. and Richards, I.G., *Use of vegetation in civil engineering*, CIRIA Book 10, 1990.

Cost effective management of reclaimed derelict sites, DoE, HMSO, 1989.

Crisp, V.H.C, Doggart, J., and Attenborough, M., BREEAM 2/91, *An environmental assessment for new superstores and supermarkets*, BRE, 1991.

Croners, *Environmental Management*, with quarterly amendment service, Croner Publications Ltd, Kingston upon Thames, First Edn, October 1991.

Curwell, S.R., March, C.G. and Venables, R.K., (Eds), *Buildings and Health, The Rosehaugh Guide to the Design, Construction, Use and Management of Buildings*, RIBA Publications, 1990.

DoE, *The use of halons in the United Kingdom and the scope for substitution*, DOE,, HMSO, 1991.

Department of Trade and Industry, *CFCs and Halons, Alternatives and the scope for recovery for recycling and destruction*, DTI, HMSO, 1990.

Elkington, J., Knight, P., Hailes, J., *The Green Business Guide – How to take up – and profit from – the environmental challenge*, Victor Gollancz, London 1991.

Energy Efficiency Office, *Energy consumption in offices – a technical guide for owners and single tenants*, Energy Consumption Guide 19, EEO, 1991.

Engineering Council, *Engineers and the Environment: Code of Professional Practice*, 1993.

Environment Council, *Business and Environment Programme Handbook*, 1992 plus regular updates.

Environmental Assessment: A guide to the identification, evaluation and mitigation of environmental issues in construction schemes, CIRIA Special Publication 96, 1993.

European Construction Institute, *Total project management of construction safety, health and environment*, Thomas Telford, 1992.

Fox, A. and Murrell, R., *Green Design: A guide to the environmental impact of building materials*, Architecture Design and Technology Press, 1989.

Government White Paper, *This Common Inheritance – Britain's Environmental Strategy (1990)* and *Yearly Reports* (1991 and 1992).

Guidance on the sale and transfer of contaminated land, Draft for open consultation, CIRIA, October 1993.

Hall, K and Warm, P., *Greener Building Products and Services Directory*, Association for Environment Conscious Building Directory, Second Edition, 1993.

Halliday, S.P., *Building Services and Environmental Issues – The Background*, BSRIA Interim Report, April 1992.

Halliday, S.P., *Environmental Code of Practice for Buildings and Their Services*, BSRIA, 1994.

HSC, *Proposals for Construction (Design and Management) Regulations and Approved Code of Practice*, HSE, 1992.
Hooker, P.J. and Bannon, M.P., *Methane: its occurrence and hazards in construction*, CIRIA Report 130, 1993.

Householder's Guide to Radon, 3rd Edition, DoE, 1992.

ICRCL 17/78, *Notes of the development and after-use of landfill sites*, ICRCL, 1990.

Johnston, J. and Newton J., *Building Green – A guide to using plants on roofs, walls and pavements*, London Ecology Unit, 1993.

Leach, B.A., and Goodger, H.K., *Building on derelict land*, CIRIA Special Publication 78, 1991.

Building for the Environment: An Environmental Good Practice Checklist for the Construction and Development Industries, Leicester County Council jointly with Leicester City Council, November 1992.

Miller, S., *Going Green*, JT Design Build, Bristol.

Ministry of Agriculture, Fisheries and Food, *Environmental Procedures for Inland Flood Defence Works* – A guide for managers and decision makers in the NRA, Internal Drainage Boards and Local Authorities, 1992.

NRA, Pollution Prevention Guidelines Working at Demolition and Construction Sites, NRA, July 1992

Ove Arup & Partners, *The Green Construction Handbook – A Manual for Clients and Construction Professionals*, JT Design Build, Bristol, 1993.

Potter, I.N., *Sick Building Syndrome*, Report BTN04/88, BSRIA, 1988.

Prior, J.J., (Editor), BREEAM 1/93, *An environmental assessment for new offices*, BRE, 1993.

Remedial treatment of contaminated land, in 12 volumes, forthcoming CIRIA publications due to be published 1994.

Skoyles, E.R. and Skoyles, J.R., *Waste Prevention on Site*, Mitchell Publishing, 1987.

Steeds, J.E., Shepherd, E. and Barry, D.L., *A guide to safe working practices for contaminated sites*, Unpublished CIRIA Core Programme Funders Report, FR/CP/9, July 1993, (in preparation as an open publication).

The Good Wood Manual: Specifying Alternatives to Non-renewable Tropical Hardwoods, Friends of the Earth, 1990.

Timber: Types and sources, Publication L296, Friends of the Earth, 1993.

Venables, R.K et al, *Environmental Handbook for Building and Civil Engineering Projects, Volume 1: Design and specification*, CIRIA Special Publication 97, 1994.

B Main legislation.

UK

Alkali Works etc Regulation Act 1906.

Building (Scotland) Act 1970.

Building Act 1984.

Building Regulations, 1991.

Clean Air Acts 1956 and 1958.

Control of Pollution Act 1974.

Control of Substances Hazardous to Health Regulations 1988 (SI 1988 No. 1657) (as amended by SI 1990 No. 2026 and SI 1992 No. 2382).

Construction Products Regulations 1991 (SI 1991 No. 1620).

Environmental Protection Act 1990.

Harbour Works (Assessment of Environmental Effects) Regulations 1988 (SI 1988 No. 1336) and (No.2) Regulations 1989 (SI 1989 No. 524).

Health and Safety at Work etc Act 1974.

Highways (Assessment of Environmental Effects) Regulations 1988 (SI 1988 No. 1241).

Land Drainage Improvement Works (Assessment of Environmental Effects) Regulations 1988 (SI 1988 No. 1217).

National Parks and Access to the Countryside Act 1949.

Noise at Work Regulations 1989 (SI 1989 No. 1790).

Occupier's Liability Acts 1957 and 1984.

Planning (Consequential Provisions) Act 1990.

Planning (Hazardous Substances) Act 1990.

Planning and Compensation Act 1991.

The Electricity and Pipe-line Works (Assessment of Environmental Effects) Regulations 1989 (SI 1989 No. 167).

Town and Country (Assessment of Environmental Effects) Regulations 1988 (SI 1988 No. 1199) (as amended by SI 1990 No. 367 and SI 1992 No. 1494).

Town and Country Planning (Scotland) Act 1972.

Town and Country Planning Act 1990.

Water Industry Act 1991.

Water Resources Act 1991.

Wildlife and Countryside Act 1981 (as amended).

Workplace (Health, Safety and Welfare) Regulations 1992 (SI 1992 No 3004).

EC

Directive on Construction Products 89/109/EEC.

Directive on Marketing and Use of Certain Dangerous Substances 76/769/EEC (as Directive on the assessment of the effects of certain public and private projects on the environment 85/337/EEC.

Directive on the Conservation of Wild Birds (79/409/EEC).

Directive on the Conservation of Wild Fauna and Flora (92/43/EEC) (Habitats Directive) (Intended to be implemented by 05.06.1994).

Directive on the minimum safety and health requirements for the workplace (89/655/EEC).

EC Regulation (880/92) on a Community Eco-Label Award Scheme.

Addresses of relevant organisations

Association of Consulting Engineers
Alliance House
12 Caxton Street
LONDON SW1H 0QL

Phone: 071-222 6557   Fax: 071-222 0750

The Association for Environment Conscious Building
Windlake House
The Pump Field
COALEY
Gloucestershire GL11 5DX

Phone: 0453 890757

Arboricultural Advisory and Information Service
Forestry Commission Research Station
Alice Holt Lodge
Wrecclesham
FARNHAM
Surrey GU10 4LH

Phone: 0794 68717

Arboricultural Association
Ampfield House
Ampfield
ROMSEY
Hampshire SO5 9PA

Phone: 0794 68717

Architectural Cladding Association
60 Charles Street
LEICESTER
Leicestershire LE1 1FB

Phone: 0533 536161   Fax: 0533 514568

Association of British Plywood and Veneer Manufacturers
Riverside Industrial Estate, Morson Road
Ponders End
ENFIELD
Middlesex EN3 4TS

Phone: 081-804 2424

Association of Building Component Manufacturers
61-63 Rochester Road
AYLESFORD
Kent ME20 7BS

Phone: 0622 715577

British Aggregate Construction Materials Industries
156 Buckingham Palace Road
LONDON SW1W 9TR

Phone: 071-730 8194   Fax: 071-730 4355

Environment Task Group
Building Employers Confederation
82 New Cavendish Street
LONDON W1M 8AD

Phone: 071-580 5588   Fax: 071-631 3872

British Association of Landscape Industries
Landscape House
9 Henry Street
KEIGHLEY
West Yorkshire BD21 3DR

Phone: 0535 606139

British Earth Sheltering Association
Caer Llan Barn House
Lyddant
MONMOUTH NP5 4JJ

British Library Environmental Information Service
25 Southampton Buildings
Chancery Lane
LONDON WC2A 1AW

Phone: 071-323 7955   Fax: 071-323 7954

British Timber Merchants' Association (BTMA)
Stocking Lane
Hughenden Valley
HIGH WYCOMBE HP14 4JZ

Phone: 0494 563602

British Standards Institution
2 Park Street
LONDON W1A 2BS

Phone: 071-629 9000

Enquiry and Ordering Department
British Standards Institution
Linford Wood
MILTON KEYNES
Bucks MK14 6LE

Phone: 0908-226888   Fax: 0908-322484

British Wood Preserving Association
Building No.6
The Office Village
4 Romford Road
Stratford
LONDON E15 4EA

Phone: 081-519 2588

Building Environment Performance Analysis Club
Building Research Establishment
Garston
WATFORD WD2 7JR

Phone: 0923 664138   Fax: 0923 664780

Building Research Energy Conservation Support Unit (BRECSU)
Building Energy Efficiency Division
Building Research Establishment
Garston
WATFORD WD2 7JR

Phone: 0923 664258   Fax: 0923 664787

Addresses of relevant organisations

Energy and Environment Group
Building Services Research and Information Association
Old Bracknell Lane West
BRACKNELL
Berkshire RG12 4AH

Phone: 0344 426511   Fax: 0344 487575

Environmental Assessment and Futures Section
Building Research Establishment
Garston
WATFORD
Herts WD2 7JR

Phone: 0923 664174 DL   Fax: 0923 664088

Business in the Environment
8 Stratton Street
LONDON
W1X 5FD

Phone: 071-629 1600   Fax: 071-629 1834

Environment Management Unit
Confederation of British Industry
Centre Point
103 New Oxford Street
LONDON WC1A 1DU

Phone: 071-379 7400   Fax: 071-240 0988

Chartered Institution of Building Services Engineers
Delta House
222 Balham High Road
LONDON SW12 9/BS

Phone: 081-675 5211   Fax: 081-675 5449

Construction Industry Council
26 Store Street
LONDON WC1E 7BY

Phone: 071-637 8692

Chartered Institute of Building
Englemere
Kings Ride
ASCOT
Berks SL5 8BJ

Phone: 0344 23355   Fax: 0344 23467

CIRIA (Construction Industry Research and Information Association)
6 Storey's Gate
Westminster
LONDON SW1P 3AU

Phone: 071-222 8891   Fax: 071-222 1708

Construction Industry Environmental Forum
CIRIA
6 Storey's Gate
Westminster
LONDON SW1P 3AU

Phone: 071-222 8891   Fax: 071-222 1708

Commission of the European Communities
8 Storey's Gate
LONDON SW1P 3AT

Phone: 071-973 1992   Fax: 071-222 0900

Energy Efficiency Office
Department of the Environment
1 Palace Street
LONDON SW1E 5HE

Phone: 071-238 3094   Fax: 071-238 3733

Construction Sponsorship Directorate
Department of the Environment
Room P1/113A
2 Marsham Street
LONDON SW1P 3EB

Phone: 071-276 6728   Fax: 071-276 3826

Dry Lining and Partition Association
82 New Cavendish Street
LONDON W1M 8AH

Phone: 071-580 5588

Energy Design Advice Scheme
The Bartlett Graduate School of Architecture
Philips House
University College, London
Gower Street
LONDON WC1E 6BT

Phone: 071-916 3891

Energy Division
Department of Trade and Industry
1 Palace Street
LONDON SW1E 5HE

Phone: 071-238 3000/3370 Fax: 071-834 3771

Energy System Trade Association
P O Box 16
STROUD
Gloucestershire
GL6 9YB

Phone: 0453 886776   Fax: 0453 885226

Environmental Enquiry Point
Department of Trade and Industry
Warren Spring Laboratory
Gunnels Wood Road
STEVENAGE
Herts SG1 2BX

Phone: 0800 585794   Fax: 0438 360858

The Business and Environment Programme
The Environment Council
21 Elizabeth Street
LONDON SW1W 9RP

Phone: 071-824 8411   Fax: 071-730 9941

European Federation of Waste Management
Avenue des Nerviens 117/69
B-1040 BRUSSELS
BELGIUM

Phone: 010 32 2 732 1601   Fax: 010 32 2 734 9279

Federation of Civil Engineers Contractors
Cowdray House, 6 Portugal Street
LONDON WC2A 2HH

Phone: 071-404 4020   Fax: 071-242 0256

Fencing Contractors' Association
St John's House
WATFORD
Hertfordshire WD1 1PY

Phone: 0923 248895

Forest Stewardship Council
c/o WWF UK
Panda House
Weyside Park
Catteshall Lane
GODALMING
Surrey GU7 1XR

Phone: 0483 426444   Fax: 0483 426409

Friends of the Earth
26/28 Underwood Street
LONDON N1 7JQ

Phone: 071-490 1555

Greenpeace
30-31 Islington Green
LONDON N1 8XE

Phone: 071-354 5100

Her Majesty's Inspectorate of Pollution
Romney House,
43 Marsham Street
LONDON SW1P 3PY

Phone: 071-276 8083   Fax: 071-276 8605

Her Majesty's Stationery Office
London Bookshop
49 High Holborn
LONDON WC1V 6HB

Phone: 071-873 0011

Institution of Civil Engineers
Great George Street
LONDON SW1P 3AA

Phone: 071-222 7722

Institute of Environmental Assessment
The Old School
Fen Road
East Kirby
LINCOLNSHIRE
PE23 4DB

Phone: 0790 763613   Fax: 0790 3630

Institute for European Environmental Policy
3 Endsleigh Street
LONDON WC1H 0DD

Phone: 071-388 2117   Fax: 071-388 2826

International Union for Conservation of Nature and Natural Resources
World Conservation Monitoring Centre
219c Huntington Road
CAMBRIDGE CB3 0DL

Phone: 0223 277314

JT Design Build Ltd
Bush House
72 Prince Street
BRISTOL BS1 4HU

Phone: 0272 290651   Fax: 0272 290946

Centre for Study for Environmental Change
Lancaster University
Fylde College
LANCASTER
Lancashire LA1 4YF

Phone: 0524 65201 X 2844   Fax: 0524 846339 DL

London Waste Regulation Authority
Hampton House
20 Albert Embankment
LONDON SE1 7TT

Phone: 071-587 3000   Fax: 071-587 5258

National Council of Building Material Producers
26 Store Street
LONDON WC1E 7BT

Phone: 071-323 3770   Fax: 071-323 0307

National Energy Foundation
Rockingham Drive
Linford Wood
MILTON KEYNES
MK14 6EG

Phone: 0908 672787   Fax: 0908 662296

National House-Building Council
Buildmark House
Chiltern Avenue
AMERSHAM
Buckinghamshire HP6 5AP

Phone: 0494 434477   Fax: 0494 728521

National Radiological Protection Board
Chilton
DIDCOT
Oxfordshire OX11 ORQ

Phone: 0235 831600   Fax: 0235 833891

Addresses of relevant organisations

National Rivers Authority
Rivers House
Waterside Drive
Aztec West
Almondsbury
BRISTOL BS12 4UD

Phone: 0454 624400   Fax: 0454 624409

New Homes Environmental Group
82 New Cavendish Street
LONDON
W1M 8AD

Phone: 071-580 5588

New Homes Marketing Board
82 New Cavendish Street
LONDON
W1M 8AD

Phone: 071-580 5588   Fax: 071-323 1697

RIBA (Royal Institute of British Architects)
66 Portland Place
LONDON WIN 4AD

Phone: 071-580 5533   Fax: 071-255 1541

Royal Institution of Chartered Surveyors
12 Great George Street
Parliament Square
LONDON SW1P 3AD

Phone: 071-222 7000   Fax: 071-222 9430

The Royal Town Planning Institute
26 Portland Place
LONDON W1N 4BE

Phone: 071-636 9107   Fax: 071-323 1582

Timber Research and Development Association
Stocking Lane
Hughenden Valley
HIGH WYCOMBE
Buckinghamshire HP14 4ND

Phone: 0494 563691   Fax: 0494 565487

Forests Forever Campaign
Timber Trade Federation
Clareville House
26/27 Oxendon Street
LONDON SW1Y 4EL

Phone: 071-839 1891   Fax: 071-930 0094

UK Ecolabelling Board
7th Floor
Eastbury House
30-34 Albert Embankment
LONDON SE1 7TL

Phone: 071-820 1199   Fax: 071-820 1104

Wood Panel Products Federation
1 Hanworth Road
FELTHAM
Middlesex TW13 5AF

Phone: 081-751 6107   Fax: 081-890 2870

World Wide Fund for Nature
Panda House
Weyside Park
GODALMING
Surrey GU7 1XR

Phone: 0483 426444   Fax: 0483 426409

Environmental Checklist: Construction Phase

Project: **Project Number:**

Name: **Date:**

                                              ☑ Date   Initials

Stage 1 Tendering

1.1 An introduction to relevant legislation and policy   ☐

1.2 Identification and evaluation of contract environmental requirements   ☐

1.3 Tenderer's potential influence on the environmental aspects of the project   ☐

1.4 Environmental matters to include in a tender   ☐

Stage 2 Project planning and contract letting

2.1 Legislation and policy

    2.1.1 The Environmental Protection Act 1990   ☐

    2.1.2 Waste legislation and policy   ☐

    2.1.3 Duty of care for waste   ☐

    2.1.4 Water legislation and policy   ☐

    2.1.5 Noise legislation and policy   ☐

    2.1.6 Health and Safety at Work etc Act, 1974   ☐

    2.1.7 Control of Substances Hazardous to Health   ☐

    2.1.8 Contaminated land legislation and policy   ☐

    2.1.9 Air pollution legislation and policy   ☐

    2.1.10 The Wildlife and Countryside Act 1981   ☐

2.2 Overall purchasing policies

    2.2.1 Agreeing green purchasing policies   ☐

    2.2.2 Awareness and training   ☐

    2.2.3 Information requirements   ☐

    2.2.4 Sub-contractor management   ☐

2.3 Green management of a site

    2.3.1 Need for and implementation of an environmental management system/BS7750   ☐

    2.3.2 Awareness and training   ☐

    2.3.3 Recording of environmental performance during the construction phase   ☐

    2.3.4 Site-specific purchasing policies   ☐

    2.3.5 Transport policies including selection and maintenance of site plant and other vehicles   ☐

    2.3.6 Minimising energy use   ☐

    2.3.7 Minimising water use   ☐

    2.3.8 Waste management, storage and re-use   ☐

    2.3.9 Strategies for dealing with sensitive areas eg archaeology, nature, conservation, SSSIs   ☐

    2.3.10 Sub-contractor management   ☐

2.4 Pollution control strategies   ☐ ▭

2.5 Contaminated land

    2.5.1 Client assurances and identification of contaminated areas on site   ☐ ▭

    2.5.2 Plans to deal with any residual contamination and/or unforeseen contamination if discovered   ☐ ▭

2.6 Relations with relevant bodies and groups

    2.6.1 Appropriate consultation with NRA, planning authorities, environmental and conservation agencies   ☐ ▭

    2.6.2 Relations with site neighbours and the local public   ☐ ▭

2.7 Special considerations for design and build projects   ☐ ▭

Stage 3 Site set-up and management

3.1 Legislation and policy   ☐ ▭

3.2 Positioning, layout and planning of the site compound   ☐ ▭

3.3 Relations with site neighbours and the local public

    3.3.1 Establishing good relations with neighbours and the local public   ☐ ▭

    3.3.2 Working hours and considerate contractors schemes   ☐ ▭

3.4 Implementing green management plans on site   ☐ ▭

3.5 Implementing pollution control strategies

    3.5.1 Waste water control   ☐ ▭

    3.5.2 Disposal of other wastes   ☐ ▭

    3.5.3 Air pollution and its control, including dust and fumes   ☐ ▭

    3.5.4 Noise and vibration control   ☐ ▭

    3.5.5 Light control   ☐ ▭

3.6 Protection of sensitive areas of the site

    3.6.1 Trees, water, species, habitat and landscape features   ☐ ▭

    3.6.2 Habitat translocation and/or creation   ☐ ▭

    3.6.3 Early planting of trees and shrubs   ☐ ▭

    3.6.4 Building structure and facades   ☐ ▭

3.7 Traffic management

    3.7.1 Legislation and policy   ☐ ▭

    3.7.2 Traffic management strategies and their implementation   ☐ ▭

    3.7.3 Production on- and off-site compared   ☐ ▭

3.8 Environmental impact of temporary works   ☐ ▭

3.9 Environment in the site offices   ☐ ▭

3.10 Special considerations for civil engineering projects   ☐ ▭

Stage 4 Demolition and site clearance

4.1 Legislation and policy   ☐ ▭

4.2 Dealing with residual and/or unforeseen contamination in land
and/or buildings and works   ☐ ▭

4.3 Waste management

    4.3.1 Waste management principles at the demolition and site
clearance stage   ☐ ▭

    4.3.2 Waste water control, and oil and petrol tanks   ☐ ▭

4.4 Fires and their control   ☐ ▭

4.5 Identification and protection of existing services   ☐ ▭

4.6 Archaeology and ecology   ☐ ▭

Stage 5 Groundworks (including earthworks)

5.1 Materials and processes of potential environmental concern at this
stage   ☐ ▭

5.2 Legislation and policy   ☐ ▭

5.3 Dealing with residual and/or unforeseen contamination in land
and/or buildings and works   ☐ ▭

5.4 Temporary storage of spoil, disposal of excess spoil and importing
earthworks materials   ☐ ▭

5.5 Hydrological, archaeological and ecological considerations

    5.5.1 Water courses and site hydrology   ☐ ▭

    5.5.2 Archaeology and ecology   ☐ ▭

5.6 Geotechnical processes and ground engineering   ☐ ▭

Stage 6 Foundations

6.1 Materials and processes of potential environmental concern at this
stage   ☐ ▭

6.2 Legislation and policy   ☐ ▭

6.3 Purchasing requirements, and sourcing and transport of materials ☐ ▭

6.4 Materials

    6.4.1 Labelling schemes   ☐ ▭

    6.4.2 Handling and storage of materials   ☐ ▭

    6.4.3 Recycling of materials   ☐ ▭

    6.4.4 Disposal of spare materials, waste and containers   ☐ ▭

6.5 Dealing with residual and/or unforeseen contamination   ☐ ▭

Project: **Project Number:**

Name: **Date:**

Stage 7 Structural work for building or civil engineering

7.1 Materials and processes of potential environmental concern at this stage ☐ □━━━━┳━━━┓

7.2 Legislation and policy ☐ □━━━━┳━━━┓

7.3 Purchasing requirements, and sourcing and transport of materials ☐ □━━━━┳━━━┓

7.4 Materials

    7.4.1 Labelling schemes ☐ □━━━━┳━━━┓

    7.4.2 Handling and storage of materials ☐ □━━━━┳━━━┓

    7.4.3 Salvage and recycling of materials ☐ □━━━━┳━━━┓

    7.4.4 Particular issues in the use of timber ☐ □━━━━┳━━━┓

    7.4.5 Disposal of spare materials, waste and containers ☐ □━━━━┳━━━┓

Stage 8 Building envelope

8.1 Materials and processes of potential environmental concern at this stage ☐ □━━━━┳━━━┓

8.2 Legislation and policy ☐ □━━━━┳━━━┓

8.3 Purchasing requirements, and sourcing and transport of materials ☐ □━━━━┳━━━┓

8.4 Materials

    8.4.1 Use of timber ☐ □━━━━┳━━━┓

    8.4.2 Labelling schemes ☐ □━━━━┳━━━┓

    8.4.3 Handling and storage of materials ☐ □━━━━┳━━━┓

    8.4.4 Use of paints and solvents ☐ □━━━━┳━━━┓

    8.4.5 Use of scarce materials ☐ □━━━━┳━━━┓

    8.4.6 Use of recycled materials ☐ □━━━━┳━━━┓

    8.4.7 Exclusion of CFCs and halons ☐ □━━━━┳━━━┓

    8.4.8 Disposal of spare materials, waste and containers ☐ □━━━━┳━━━┓

Stage 9 Mechanical and electrical installations and their interface with civil and building work

9.1 Legislation and policy ☐ □━━━━┳━━━┓

9.2 Co-ordination between structural, civil and services contractors, and use of the BSRIA Environmental Code of Practice ☐ □━━━━┳━━━┓

Stage 10 Trades: joinery, painting, plastering etc

10.1 Materials and processes of potential environmental concern at this stage ☐ □━━━━┳━━━┓

10.2 Legislation and policy ☐ □━━━━┳━━━┓

10.3 Purchasing requirements, and sourcing and transport of materials ☐ □━━━━┳━━━┓

10.4 Use, handling and storage of materials and components by the finishing trades ☐ □━━━━┳━━━┓

10.5 Commissioning ☐ □━━━━┳━━━┓

Stage 11 Landscaping, reinstatement and habitat restoration or creation

11.1	Legislation and policy	☐
11.2	Purchasing requirements	☐
11.3	Planting and maintenance	
	11.3.1 Ground preparation	☐
	11.3.2 Dealing with residual contamination	☐
11.4	Landscape establishment, and habitat restoration or creation	☐

Stage 12 Site reinstatement, removal of site offices and final clear away

12.1	Legislation and policy	☐
12.2	Options for salvage and recycling	☐
12.3	Disposal of waste	☐
12.4	Reinstatement of the site	☐

Stage 13 Handover, and guidance on maintenance and records

13.1	Legislation and policy	☐
13.2	Records of materials used and treatment of contaminated land, and advice to owners and occupiers	☐

Index

For main legislation, see Part B of the main references – page 130. Further cross references are listed in the text.